Hydrostatic Pressure Chemistry

Tomoyuki Hamachi • Gaku Fukuhara

Hydrostatic Pressure Chemistry

Concept, Strategies, and Applications

Springer

Tomoyuki Hamachi
Institute for Materials Chemistry and Engineering
Kyushu University
Fukuoka, Japan

Gaku Fukuhara
Institute for Materials Chemistry and Engineering
Kyushu University
Fukuoka, Japan

ISBN 978-981-95-7460-5 ISBN 978-981-95-7461-2 (eBook)
https://doi.org/10.1007/978-981-95-7461-2

This Springer imprint is published by the registered company Springer Nature Singapore Pte Ltd.
The registered company address is: 152 Beach Road, #21-01/04 Gateway East, Singapore 189721, Singapore

Preface

Chemical reactions are governed by a delicate balance of thermodynamic and kinetic factors that determine how molecules transform, organize, and function. Among the fundamental thermodynamic variables, temperature has long been the dominant tool for controlling chemical reactivity. However, in many modern chemical systems, particularly those involving photoexcited states, weak intermolecular interactions, and collective molecular organization, temperature often proves insufficient as a predictive or selective control parameter. Excited-state processes evolve on ultrafast timescales and are highly sensitive to subtle structural and solvation changes, rendering their behavior difficult to rationalize within conventional thermodynamic frameworks.

Hydrostatic pressure provides an alternative and complementary thermodynamic variable that directly couples to molecular volume. Unlike temperature, pressure perturbs chemical systems through volumetric changes associated with conformational rearrangements, solvation, and molecular organization, while leaving chemical composition unchanged. Because reaction volume changes are often more accessible experimentally and conceptually than entropy changes, pressure offers a uniquely quantitative and reversible means of modulating chemical equilibria and reaction pathways. This feature is particularly powerful in soft and flexible systems, including molecular assemblies, supramolecular architectures, polymers, and biological materials, where small structural reorganizations can lead to pronounced functional responses.

High-pressure chemistry has a long experimental history, spanning early spectroscopic studies of molecular solutions to investigations of phase transitions, electron transfer, and molecular association. Over time, it has become clear that pressure exerts especially strong effects on photoexcited states, excimers, charge-transfer interactions, and host-guest dynamics, all of which involve significant changes in effective molar volume. Despite these observations, pressure and temperature effects were traditionally treated independently, and a unified thermodynamic description remained elusive. Recent advances have established a coherent framework that integrates thermal and pressure contributions to free energy, enabling systematic interpretation of pressure-controlled reactions across both ground and excited states.

Beyond molecular and supramolecular chemistry, hydrostatic pressure provides a natural conceptual bridge to biology. Living systems are inherently soft, hydrated, and structurally dynamic, making them highly sensitive to volumetric perturbations. In recent years, mechanobiology has revealed that mechanical stimuli play essential roles in cellular function, differentiation, and disease progression. Yet the molecular mechanisms by which cells sense and transduce mechanical forces remain incompletely understood. Hydrostatic pressure, acting isotropically and uniformly, offers a biologically compatible means of applying well-defined mechanical stress, while pressure-responsive optical probes enable direct visualization of these effects in complex biological environments.

This book summarizes recent developments in hydrostatic pressure effects on functional molecular, supramolecular, polymeric, and biological systems in solution. Chapters 1 and 2 provide readers with the fundamental principles of chemistry under hydrostatic pressure, based on basic thermodynamic equations, and give an overview of experimental apparatuses used for measurements under controlled pressure conditions. Chapter 3 describes biological systems whose structures and functions are regulated by hydrostatic pressure. Chapter 4 describes how pressure can be used to control ground states, photoexcited states, and chemical reactions. Chapter 5 focuses on supramolecular chemosensors in which molecular recognition processes are regulated by pressure. Chapter 6 gives polymer-based chemosensors that enable highly sensitive optical readout of pressure stimuli through fluorescence. Finally, Chapter 7 summarizes recent advances in hydrostatic pressure chemistry.

This book aims to provide a coherent thermodynamic perspective on pressure-controlled chemistry and biology. It is our hope that this book will offer both newcomers and experienced researchers a conceptual foundation for understanding how volumetric effects shape molecular behavior, and will stimulate further exploration of pressure as a fundamental and versatile stimulus in chemistry, materials science, and life science.

Fukuoka, Japan Tomoyuki Hamachi
Fukuoka, Japan Gaku Fukuhara

Competing Interests The authors have no competing interests to declare that are relevant to the content of this manuscript.

Contents

Chapter 1
Introduction

Abstract Controlling chemical reactions with precision is a central objective across chemistry, materials science, and biology. While thermally driven reactions are well described by classical thermodynamics and catalyst design, photochemical processes remain considerably more difficult to predict because excited states evolve on ultrafast timescales and are governed by weak intermolecular interactions. As a result, their reactivity is highly sensitive to the balance between enthalpic and entropic contributions, and temperature often fails as a reliable control variable due to enthalpy-entropy compensation and increased nonradiative decay at elevated temperatures. These limitations highlight the need for an alternative thermodynamic parameter capable of modulating reaction pathways independently of entropy. Hydrostatic pressure provides such a variable by coupling directly to reaction volume changes associated with conformational rearrangements, solvation, and molecular organization. Early studies in molecular, supramolecular, and polymer systems revealed pronounced pressure effects on excimer formation, charge transfer (CT) states, and host-guest dynamics, yet only recently have these phenomena been unified within a coherent framework based on reaction volume changes. Extending this concept to biological environments has further strengthened the connection between photochemistry and mechanobiology, enabling the development of novel chemosensors that probe how cells respond to mechanical stimuli. This chapter outlines these unified thermodynamic principles and their implications for reaction control across systems ranging from individual molecules to living systems.

Keywords Thermodynamics · Hydrostatic pressure · Photoexcited state · Mechanochemistry · Mechanobiology

1.1 Thermodynamic Framework for Photoreaction Control

A central objective of chemical science is to achieve precise control over the thermodynamics and kinetics of reactions, regardless of whether they proceed on the ground or excited states. For thermal reactions, substantial progress has been made

T. Hamachi, G. Fukuhara, *Hydrostatic Pressure Chemistry*,
https://doi.org/10.1007/978-981-95-7461-2_1

in lowering activation barriers and stabilizing transition states through the development of catalysts that accelerate reaction rates. In contrast, photochemical reactions, which began to flourish in the mid-1960s following the proposal of the Woodward–Hoffmann rules (Woodward & Hoffmann, 1969), have advanced primarily through empirical exploration of external environmental factors such as temperature, solvation, and supramolecular interactions. Because excited states typically exhibit extremely small activation energies, product distributions (ΔG) in photochemical systems are governed by a delicate balance between enthalpy (ΔH) and entropy (ΔS) contributions, as expressed by the Gibbs free energy relationship (Eq. 1.1).

$$\Delta G = \Delta H - T\Delta S \tag{1.1}$$

Equation (1.1) illustrates that reaction spontaneity depends on the interplay between ΔH and ΔS. In systems dominated by strong interactions, such as covalent or coordination bonds (~100 kcal mol^{-1}), ΔH overwhelmingly determines ΔG, and the ΔS contribution is often negligible. By contrast, in photochemical reactions governed by weak interactions (<10 kcal mol^{-1}) including electrostatic, van der Waals, hydrophobic, hydrogen-bonding, and π-π interactions, both ΔH and ΔS contribute comparably to the overall free energy change. Unlike thermal reactions, which have traditionally been interpreted mainly in terms of ΔH, the importance of ΔS has long been emphasized in photochemical processes. As Inoue et al. comprehensively reviewed (Yang & Inoue, 2014), excited-state reactions are particularly susceptible to entropic effects because photoexcitation induces dynamic structural relaxation and solvent reorganization. Nevertheless, even with advanced computational methodologies, accurate prediction of ΔS in aggregated or condensed excited-state systems remains extremely challenging.

In such weakly bound photochemical systems, the enthalpy-entropy compensation effect, first articulated by Leffler (1955), often becomes dominant. This relationship implies that endothermic enthalpic changes (positive ΔH) are accompanied by compensating entropy gains (positive ΔS), resulting in a nearly constant Gibbs free energy (ΔG). Consequently, temperature perturbations typically induce simultaneous and opposing changes in ΔH and ΔS. Although heating may enhance entropy, it often destabilizes favorable enthalpic interactions, leading to only minimal shifts in equilibrium. Furthermore, higher temperatures activate additional molecular vibrational and rotational modes in photoexcited states, enhancing nonradiative decay pathways. Taken together, these effects render temperature an inefficient and often nonselective variable for predictably steering reaction equilibria.

The intrinsic difficulty of quantifying entropy, combined with the enthalpy-entropy compensation effect, renders the rational design of photochemical reactivity based solely on classical thermodynamic principles highly challenging. This inherent limitation underscores the need for a new conceptual framework that enables systematic and predictable control of photochemical reactions.

1.2 Pressure as an Orthogonal Thermodynamic Parameter

To overcome these limitations, attention has been directed toward pressure (P), a fundamental thermodynamic variable, that couples directly to reaction volume changes (ΔV) according to Eq. (1.2).

$$\Delta G = \Delta F + P\Delta V \tag{1.2}$$

where ΔF indicates the change in Helmholtz free energy. ΔV reflects tangible molecular events such as conformational rearrangements, solvation, and cavity formation during molecular recognition. Depending on whether these processes involve net volume expansion or contraction, ΔV may be positive or negative, enabling pressure to shift equilibria and alter reaction rates accordingly. Importantly, ΔV is often more directly measurable and more predictable than ΔS, making pressure a powerful thermodynamic parameter orthogonal to temperature for reaction control without altering the chemical composition of the system. Hydrostatic pressure is particularly well suited for flexible molecular, soft matter, supramolecular, polymeric, and biological systems, offering an isotropic, reversible, and quantitative means of mapping and manipulating their thermodynamic landscapes.

High-pressure chemistry has a long experimental history. As early as the 1960s, the effects of hydrostatic pressure on molecules in homogeneous solutions were investigated spectroscopically (Bovey & Yanari, 1960; Robertson, 1960). In the 1980s, pressure was shown to modulate both ground-state and excited-state processes, including intramolecular electron transfer (Rollinson & Drickamer, 1980; Hara et al., 1984), excimer and exciplex formation (Hara & Yano, 1988; Rettig et al., 1994), and twisted intramolecular CT (TICT) (Hara et al., 1988; Hara & Rettig, 1992), demonstrating the pronounced sensitivity of excited states to ΔV. By the 1990s, these insights converged with supramolecular chemistry: host–guest systems such as crown ethers (Letcher et al., 1994), cyclodextrins (Saudan et al., 2001), calixarenes (Isaacs et al., 1997), and related architectures exhibited large negative ΔV values (up to ~100 cm^3 mol^{-1}) upon guest complexation, emphasizing the mechanistic importance of reaction volume changes.

1.3 Unified Thermodynamic Framework for Pressure Control

Despite these findings, thermal and pressure effects were historically treated independently; Eqs. (1.1) and (1.2) were rarely considered within a unified thermodynamic framework. A conceptual turning point emerged when Fukuhara et al. proposed an integrated interpretation combining both equations (Mizuno & Fukuhara, 2022). They demonstrated that pressure serves as a universal external parameter capable of controlling reaction equilibria and kinetics across ground and excited states through ΔV, while minimizing the unpredictable influence of ΔS. Because ΔV originates from structural and solvation changes that can be both computationally estimated and experimentally measured, pressure offers a quantitatively accessible control variable. This is particularly significant for aggregated or collective systems, such as excitonic assemblies where the total reaction volume is a decisive factor. In such systems, hydrostatic pressure enables dynamic modulation of chemical and photochemical pathways. This unified perspective represents a genuine paradigm shift in reaction control, bridging thermodynamics, kinetics, and photophysics.

Consequently, pressure-controlled photochemistry has emerged as a frontier field linking ground- and excited-state processes. From early studies of excimer/exciplex formation to recent demonstrations of pressure-tunable CT emission and host-guest photodynamics, hydrostatic pressure has proven capable of finely adjusting molecular geometries, electronic coupling, and reaction rates. Within this framework, pressure complements—and in some cases surpasses—temperature as a practical and predictive control parameter.

1.4 Extension from Molecular to Biological Systems

Beyond molecular reactions, pressure provides a natural bridge between chemistry and biology. The concept of mechanobiology, which explores how living cells perceive and respond to mechanical forces, has attracted substantial attention in recent years. The origin of this field can be traced back to D'Arcy Thompson's *On Growth and Form* (Thomson, 1917), which proposed that biological morphology is fundamentally linked to mechanical stress. However, quantitative investigation became possible only after the invention of the atomic force microscope (AFM) by Binnig et al. (1986) and optical tweezers by Ashkin (1970), which allowed direct force measurements and cell manipulation at the microscale.

Subsequent studies quantified intercellular adhesion forces (Florin et al., 1994), the Young's modulus of cancer cell lines (Lekka et al., 1999), and the elasticity of tumor tissues (Wang et al., 2012), establishing that malignant tissues exhibit lower stiffness and that elasticity correlates with cancer progression. In 2006, Engler et al. demonstrated that the differentiation of mesenchymal stem cells (MSCs) depends on substrate elasticity (Engler et al., 2006). Soft matrices induce neuronal differentiation, intermediate stiffness leads to myogenic differentiation, and rigid matrices promote osteogenesis. These discoveries established the modern foundation of mechanobiology, wherein cells sense the mechanical properties of their environment and transduce them into biochemical signals that regulate gene expression.

Nevertheless, it remains largely unclear which cellular components act as mechanosensors and how physical forces are converted into biological responses. To address this challenge, chemosensors capable of optically visualizing local pressure distributions within living cells and tissues are highly required. Conventional mechanobiological approaches typically employ dynamic pressures such as ultrasound or shock waves that are anisotropic and generate spatial pressure gradients, thereby complicating mechanistic interpretation. In contrast, hydrostatic pressure is isotropic and uniformly transmitted, providing a controllable, quantitative, and biologically compatible mechanical stimulus. In recent years, chemosensors exhibiting optical responses to hydrostatic pressure have been developed, offering new opportunities for visualizing pressure distributions in living systems.

1.5 Scope of This Book

Throughout this book, recent advances on hydrostatic-pressure effects in functional molecular, supramolecular, polymeric, and biological systems in solution are summarized. The design and operation of optical high-pressure apparatuses that enable precise spectroscopic measurements under controlled hydrostatic pressure conditions are first described. Representative examples of pressure-dependent photochemical and biological phenomena are then discussed, including live-cell experiments performed under pressure and the development of luminescent chemosensors capable of visualizing local pressure variations. Together, these studies demonstrate how pressure can serve as a primary and unifying control parameter that links thermodynamics, photochemistry, and mechanobiology.

References

Ashkin, A. (1970). Acceleration and trapping of particles by radiation pressure. *Physical Review Letters, 24*, 156–159.

Binnig, G., Quate, C. F., & Gerber, C. (1986). Atomic force microscope. *Physical Review Letters, 56*, 930–933.

Bovey, F. A., & Yanari, S. S. (1960). Effect of solvent polarizability on the ultra-violet spectral shifts of aromatic compounds. *Nature, 186*, 1042–1044.

Engler, A. J., Sen, S., Sweeney, H. L., & Discher, D. E. (2006). Matrix elasticity directs stem cell lineage specification. *Cell, 126*, 677–689.

Florin, E.-L., Moy, V. T., & Gaub, H. E. (1994). Adhesion forces between individual ligand-receptor pairs. *Science, 264*, 415–417.

Hara, K., & Rettig, W. (1992). Effect of pressure on the fluorescence of TICT states in (*N,N*-dimethylamino)benzonitrile and its related compounds. *The Journal of Physical Chemistry, 96*, 8307–8309.

Hara, K., & Yano, H. (1988). High-pressure study on intramolecular excimer formation of 1,3-di-1-pyrenylpropane in various solvents. *Journal of the American Chemical Society, 110*, 1911–1915.

Hara, K., Arase, T., & Osugi, J. (1984). Effect of high pressure on intramolecular electron-transfer luminescence of 9,9′-bianthryl in different solvents. *Journal of the American Chemical Society, 106*, 1968–1972.

Hara, K., Suzuki, H., & Rettig, W. (1988). A high-pressure study of fluorescence peak shifts for ^1L_b, ^1L_a, and TICT states. *Chemical Physics Letters, 145*, 269–272.

Isaacs, N. S., Nichols, P. J., Raston, C. L., Sandova, C. A., & Young, D. J. (1997). Solution volume studies of a deep cavity inclusion complex of C_{60}: [*p*-benzylcalix[5]arene ⊂ C_{60}]. *Chemical Communications, 9*, 1839–1840.

Leffler, J. E. (1955). THE ENTHALPY-ENTROPY RELATIONSHIP AND ITS IMPLICATIONS FOR ORGANIC CHEMISTRY. *The Journal of Organic Chemistry, 20*, 1202–1231.

Lekka, M., Laidler, P., Gil, D., Lekki, J., Stachura, Z., & Hrynkiewicz, A. Z. (1999). Elasticity of normal and cancerous human bladder cells studied by scanning force microscopy. *European Biophysics Journal, 28*, 312–316.

Letcher, T. M., Mercer-Chalmers, J. D., & Kay, R. L. (1994). Volume changes in complex formation between crown ethers of cryptand-222 and alkali metals in various solvents. *Pure and Applied Chemistry, 66*, 419–427.

Mizuno, H., & Fukuhara, G. (2022). Solution-state hydrostatic pressure chemistry: Application to molecular, supramolecular, polymer, and biological systems. *Accounts of Chemical Research, 55*, 1748–1762.

Rettig, W., Gilabert, E., & Rulliére, C. (1994). Pressure dependence of bicimer formation in 4-dimethylamino-4′-cyanostilbene and model compounds. *Chemical Physics Letters, 229*, 127–133.

Robertson, W. W. (1960). Comparison of the effects of high pressure and low temperature on the absorption spectra of some condensed-ring aromatics. *The Journal of Chemical Physics, 33*, 362–365.

Rollinson, A. M., & Drickamer, H. G. (1980). High pressure study of luminescence from intramolecular CT compounds. *The Journal of Chemical Physics, 73*, 5981–5996.

Saudan, C., Dunand, F. A., Abou-Hamdan, A., Bugnon, P., Lye, P. G., Lincoln, S. F., & Merbach, A. E. (2001). A model for sequential threading of α-cyclodextrin onto a guest: A complete thermodynamic and kinetic study in water. *Journal of the American Chemical Society, 123*, 10290–10298.

Thomson, J. A. (1917). On growth and form. *Nature, 100*, 21–22.

Wang, S., Li, J., Manapuram, R. K., Menodiado, F. M., Ingram, D. R., Twa, M. D., Lazar, A. J., Lev, D. C., Pollock, R. E., & Larin, K. V. (2012). Noncontact measurement of elasticity for the detection of soft-tissue tumors using phase-sensitive optical coherence tomography combined with a focused air-puff system. *Optics Letters, 37*, 5184–5186.

Woodward, R. B., & Hoffmann, R. (1969). The conservation of orbital symmetry. *Angewandte Chemie, International Edition, 8*, 781–853.

Yang, C., & Inoue, Y. (2014). Supramolecular photochirogenesis. *Chemical Society Reviews, 43*, 4123–4143.

Chapter 2
Hydrostatic Pressure Apparatus

Abstract Hydrostatic pressure spectroscopy is carried out using a custom-built optical high-pressure apparatus that enables diverse spectroscopic measurements under precisely controlled isotropic pressure conditions. The system consists of two primary components: a quartz-based inner cell that contains the sample solution and a stainless-steel outer pressure cell that transmits hydraulic pressure. The inner cell, composed of a quartz cuvette, a Teflon tube, and a metallic cap, withstands pressures up to approximately 500 MPa and is hermetically sealed before being placed inside the water-filled outer cell. Pressure is applied through a hand-operated pumping unit, which delivers uniform hydrostatic pressure to the sample by compressing the surrounding water. The outer pressure cell is equipped with sapphire or quartz windows for UV-vis absorption, fluorescence, excitation, and time-correlated single-photon counting measurements, while diamond or YAG windows are employed for circular dichroism (CD) spectroscopy to avoid optical artifacts such as linear dichroism and birefringence. The pressure cell can be readily mounted on a range of spectroscopic instruments, enabling stable and reproducible acquisition of optical data. This design provides a versatile platform for systematically probing pressure-dependent photophysical and photochemical processes in molecular, supramolecular, polymeric, and biological systems under fully isotropic hydrostatic pressure conditions.

Keywords Hydrostatic pressure spectroscopy · High-pressure cell

Spectroscopic measurements under hydrostatic pressure are conducted using a custom-built optical high-pressure spectroscopic system (Fig. 2.1) (Mizuno et al., 2020; Kinoshita et al., 2023), specifically designed to enable a wide range of spectroscopic measurements under precisely controlled hydrostatic conditions (Ayitou et al., 2013).

The setup consists of two main components: an inner cell that contains the sample solution and an outer pressure cell made of stainless steel. The inner cell (Fig. 2.1a) is composed of a quartz cuvette (3 mm × 2 mm × 7 mm), a Teflon tube, and a metallic cap, and it can withstand hydrostatic pressures up to approximately 500 MPa. The sample solution is introduced into the quartz cell through the Teflon

T. Hamachi, G. Fukuhara, *Hydrostatic Pressure Chemistry*,
https://doi.org/10.1007/978-981-95-7461-2_2

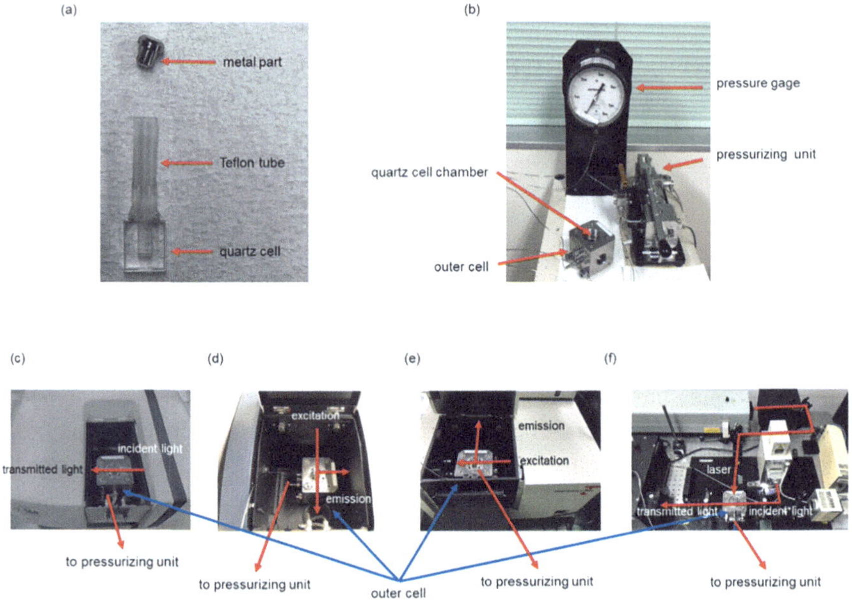

Fig. 2.1 Photographs of the (**a**) inner quartz cell, (**b**) pressurizing units, and experimental setups for (**c**) UV-vis absorption, (**d**) fluorescence/excitation, (**e**) time-correlated single-photon counting, and (**f**) nanosecond transient absorption (nsTA) measurements. (Reproduced with permission. Copyright 2023 Royal Society of Chemistry)

tube, after which the cell is hermetically sealed with the metallic cap. The sealed inner cell is then placed in a cell holder inside the outer pressure cell (Fig. 2.1b), which is completely filled with water and tightly closed.

Pressure is applied using a hand-operated pumping unit connected to the outer cell. As water is pumped in, the transmitted hydraulic pressure acts isotropically on the Teflon tube of the inner cell, thereby applying uniform hydrostatic pressure to the sample solution. After pressure stabilization, the outer cell can be mounted onto various spectroscopic instruments to perform measurements under controlled pressure.

The stainless-steel outer cell is equipped with three optical windows. Sapphire or quartz windows are used for UV-vis absorption, fluorescence/excitation, and time-correlated single-photon counting (TCSPC) measurements, whereas diamond or YAG windows are used for CD spectroscopy. The use of YAG windows in CD measurements is essential for eliminating spectral artifacts such as linear dichroism and birefringence that can arise from window-induced optical anisotropy. Figure 2.1c–f show the configurations of the pressure cell when mounted on UV-vis, fluorescence/excitation, TCSPC, and nsTA instruments, respectively.

This design allows for stable and reproducible optical measurements of molecular and supramolecular systems under isotropic hydrostatic pressure, thereby enabling systematic investigation of pressure-dependent photophysical and photochemical processes.

References

Ayitou, A. J.-L., Fukuhara, G., Kumarasamy, E., Inoue, Y., & Sivaguru, J. (2013). Enantiospecific photochemical transformations under elevated pressure. *Chemistry—A European Journal, 19*, 4327–4334.

Kinoshita, T., Nakamura, S., Harada, M., Hasobe, T., & Fukuhara, G. (2023). Control of intramolecular singlet fission in a pentacene dimer by hydrostatic pressure. *Chemical Science, 14*, 3293–3301.

Mizuno, H., Kitamatsu, M., Imai, Y., & Fukuhara, G. (2020). Smart fluorescence materials that are controllable by hydrostatic pressure: Peptide–pyrene conjugates. *ChemPhotoChem, 4*, 502–507.

Chapter 3
Hydrostatic Pressure-Regulated Biological Systems

Abstract Hydrostatic pressure provides an isotropic mechanical stimulus capable of modulating biological equilibria and reaction rates through volumetric effects. To explore how pressure perturbs signaling processes in living cells, HeLa cells were employed as a model system and intracellular Ca^{2+} dynamics were monitored using Fluo-4 AM under stepwise pressurization and depressurization. Control experiments in buffered solution revealed that Fluo-4 AM fluorescence decreases monotonically with increasing pressure due to pressure-promoted Ca^{2+} dissociation, while DAPI, used as an in situ indicator of cell viability, showed negligible pressure dependence. In contrast, cell suspensions exhibited a distinct response. Pressurization produced little immediate change, whereas depressurization triggered rapid and reproducible increases in Fluo-4 AM fluorescence, indicative of transient cytosolic Ca^{2+} elevations. Time-resolved traces showed gradual decay of this signal, reflecting cellular homeostatic mechanisms such as Ca^{2+} resequestration or efflux. Pharmacological depletion of endoplasmic reticulum stores with thapsigargin attenuated but did not abolish depressurization-induced responses, and removal of extracellular Ca^{2+} via EGTA did not suppress the fluorescence increase. These findings implicate intracellular Ca^{2+} release as the primary source of the pressure-regulated signal. Furthermore, this study demonstrates that hydrostatic pressure offers a quantitative means to modulate intracellular signaling pathways and establishes a foundation for mechanistic studies of mechanotransduction in living cells.

Keywords Hydrostatic pressure · Mechanobiology · Mechanotransduction · Intracellular Ca^{2+} signaling

Controlling biological equilibria and rates by external stimuli such as temperature, pH, solvent composition, and electronic excitation is essential for both fundamental investigations and technological applications. Hydrostatic pressure, an isotropic mechanical stimulus that enters thermodynamic expressions through a volumetric term, offers a compelling route to modulate equilibria and kinetics independently of enthalpy-entropy compensation effects that often frustrate precise control of weakly interacting systems. Extending pressure control to live cells therefore promises new

T. Hamachi, G. Fukuhara, *Hydrostatic Pressure Chemistry*,
https://doi.org/10.1007/978-981-95-7461-2_3

mechanistic insight into mechanotransduction phenomena as well as practical strategies for regulating cellular processes. HeLa cells were selected as a model living system to probe pressure-regulated calcium dynamics, and Fluo-4 AM, a Ca^{2+}-sensitive fluorescent probe (Schneidereit et al., 2016), was used to monitor intracellular calcium responses under stepwise pressurization and depressurization up to 300 MPa (Fig. 3.1) (Fukuchi et al., 2021). DAPI fluorescence was simultaneously monitored as an in situ indicator of cell membrane integrity and viability (Fig. 3.1).

Instrumental methods employed a custom-built high-pressure spectroscopic apparatus coupled to steady-state and confocal fluorescence detection. HeLa cells were cultured, harvested, and loaded with Fluo-4 AM and DAPI in Leibovitz's L-15 medium prior to placement in a high-pressure cuvette. Control experiments first characterized the intrinsic pressure response of Fluo-4 AM in a buffer containing $CaCl_2$ and confirmed that the Fluo-4 AM fluorescence decreases monotonically with increasing pressure (Fig. 3.2a), consistent with the pressure-induced dissociation of Ca^{2+} from the dye arising from a positive binding volume change (Schneidereit et al., 2016). In contrast, DAPI fluorescence exhibited negligible pressure

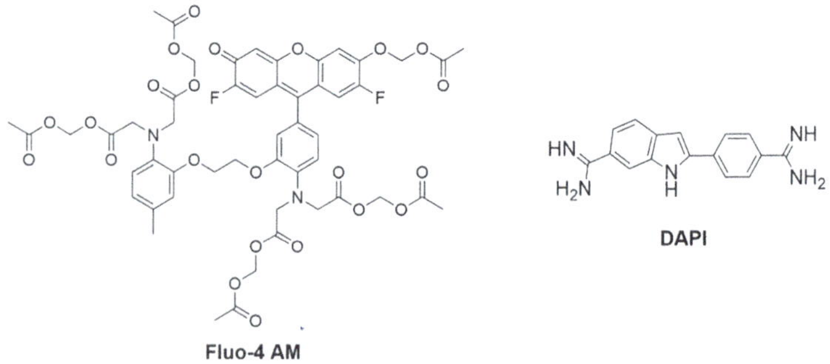

Fig. 3.1 Molecular structures of Fluo-4 AM (left) and DAPI (right)

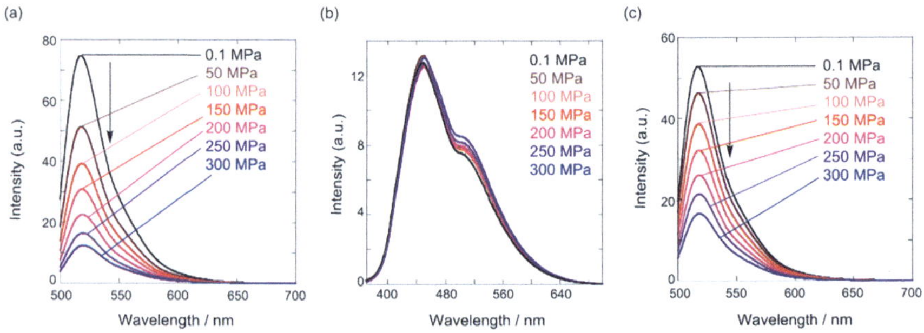

Fig. 3.2 Fluorescence spectra of (**a**) Fluo-4 AM (1 μM; λ_{ex} = 490 nm), (**b**) DAPI (3 μM; λ_{ex} = 360 nm), and (**c**) a mixture of Fluo-4 AM (1 μM) and DAPI (3 μM) (λ_{ex} = 490 nm), measured at 0.1, 50, 100, 150, 200, 250, and 300 MPa (from black to blue lines) in an L-15 medium at room temperature using a high-pressure cuvette. (Reproduced with permission. Copyright 2021 American Chemical Society)

dependence under identical conditions, validating its use as a viability marker that does not spectrally interfere with Fluo-4 AM excitation or emission (Fig. 3.2b, c).

Spectroscopic measurements of HeLa cell suspensions revealed markedly different behavior from that observed for Fluo-4 AM in buffer. At ambient pressure (0.1 MPa), Fluo-4 AM fluorescence from cell suspensions was extremely weak (Fig. 3.3a), consistent with low resting cytosolic Ca^{2+} concentrations (~100 nM) and effective dye compartmentalization. Application of 300 MPa resulted in little immediate fluorescence change while under pressure, whereas depressurization back to 0.1 MPa produced a rapid and pronounced increase in Fluo-4 AM emission intensity. Subsequent pressurization/depressurization cycles produced reproducible but attenuated fluorescence responses. In situ DAPI monitoring during these cycles showed only minor and noncritical changes in emission intensity (Fig. 3.3b), indicating that membrane integrity was largely preserved during pressure cycling as measured by the fluorescence assay. Severe membrane disruption by hydrostatic pressure would have resulted in an immediate increase in DAPI fluorescence, as

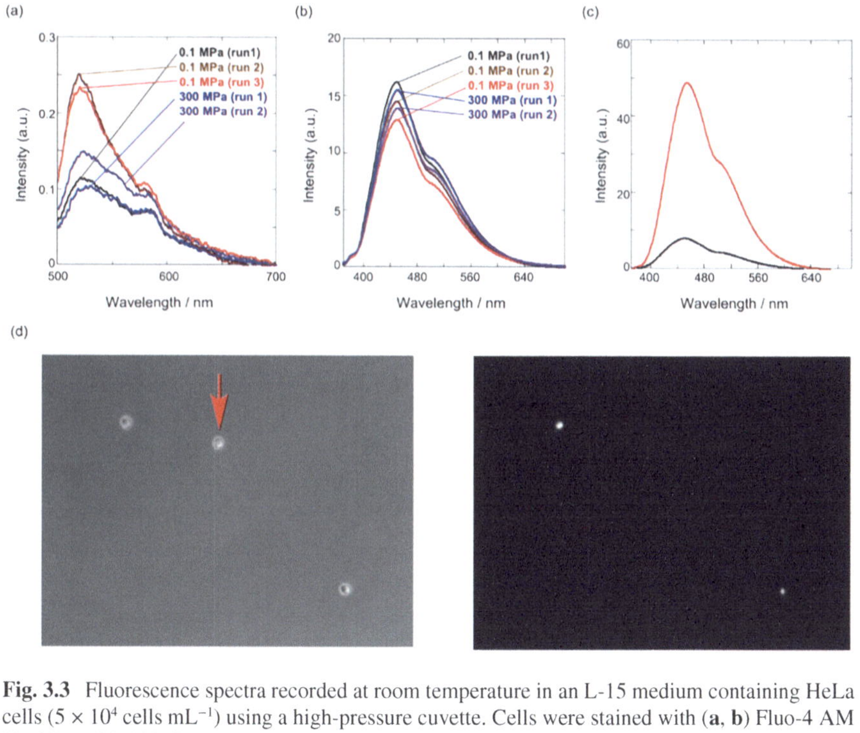

Fig. 3.3 Fluorescence spectra recorded at room temperature in an L-15 medium containing HeLa cells (5×10^4 cells mL^{-1}) using a high-pressure cuvette. Cells were stained with (**a**, **b**) Fluo-4 AM (1 μM) and DAPI (3 μM) (λ_{ex} = 490 nm), or (**c**) DAPI alone (3 μM) in the absence (black curve) and presence (red curve) of Triton X-100 (1 vol%) (λ_{ex} = 360 nm). In panels (**a**) and (**b**), pressures were applied sequentially as follows: 0.1 MPa (run 1, black) → 300 MPa (run 1, blue) → 0.1 MPa (run 2, brown) → 300 MPa (run 2, purple) → 0.1 MPa (run 3, red). Panel (d) shows phase-contrast (left) and fluorescence (right; λ_{ex} = 360 nm, λ_{obs} = 460 nm) images of HeLa cells after the hydrostatic pressure-fluorescence spectroscopy experiments, with the red arrow indicating a live cell. (Reproduced with permission. Copyright 2021 American Chemical Society)

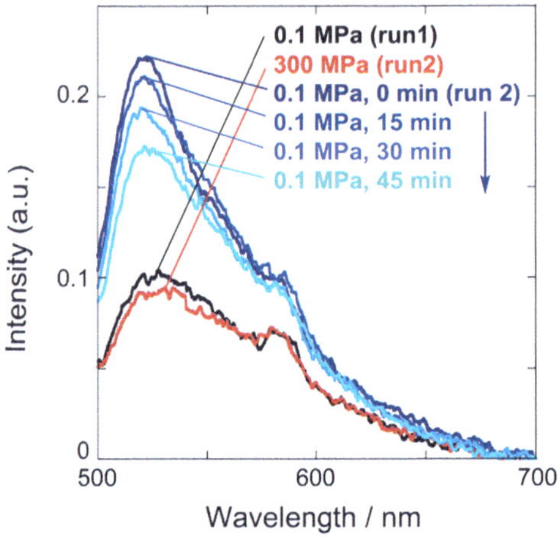

Fig. 3.4 Time-course fluorescence spectra of HeLa cell suspension in an L-15 medium (5×10^4 cells mL^{-1}) stained with Fluo-4 AM (1 μM) and DAPI (3 μM; $\lambda_{ex} = 490$ nm) recorded at room temperature using a high-pressure cuvette. The applied pressures followed the sequence: 0.1 MPa (run 1, black) → 300 MPa (run 2, red) → 0.1 MPa (run 2; 0, 15, 30, and 45 min, from blue to sky-blue curves). (Reproduced with permission. Copyright 2021 American Chemical Society)

observed after Triton X-100 treatment (Fig. 3.3c). In contrast, fluorescence microscopy performed several hours after pressurization revealed 96% cell death (Fig. 3.3d), which is attributed to the serious mechanical damage caused by ejection of the sample solution from the narrow Teflon tube of the pressure cuvette. Consistent with this interpretation, syringe-induced mechanical stimulation produced up to 59% cell death, demonstrating that such ejection-related mechanical stress can significantly compromise cell viability. Therefore, in situ fluorescence spectroscopy using DAPI provides a more reliable means of assessing pressure-induced cell death than direct fluorescence microscopy.

Time-course fluorescence traces recorded after the depressurization-induced signal jump displayed a gradual decay of tens of minutes, suggesting the activation of cellular homeostatic mechanisms that restore basal cytosolic Ca^{2+} levels (Fig. 3.4). These kinetics are consistent with transient opening of ion channels followed by active resequestration of Ca^{2+} into intracellular stores or extrusion across the plasma membrane, thereby returning the system toward baseline.

The origin of the Ca^{2+} signal was further dissected by distinguishing intracellular versus extracellular sources using pharmacological and chelation approaches. Thapsigargin was used to deplete endoplasmic reticulum (ER) Ca^{2+} stores by inhibiting the sarco-/endoplasmic reticular Ca^{2+}-ATPase (SERCA) (Schneidereit et al., 2016), while extracellular Ca^{2+} was removed from the medium using EGTA (Hotta et al., 2000, Wei et al., 2009). Under thapsigargin treatment, depressurization continued to elicit a sudden fluorescence increase (Fig. 3.5a), although with reduced amplitude relative to untreated cells. In contrast, chelation of extracellular Ca^{2+} with

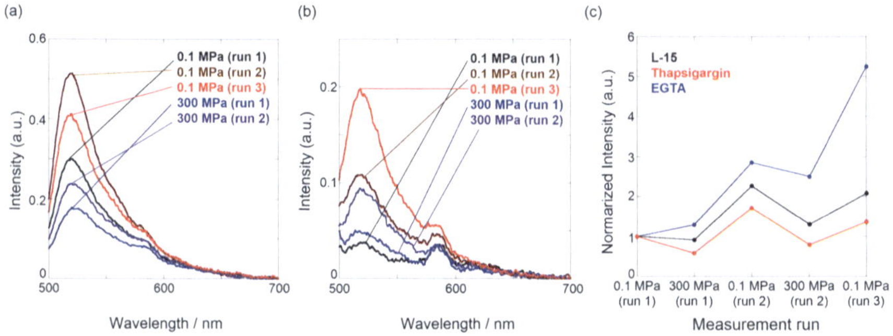

Fig. 3.5 Fluorescence spectra of HeLa cell suspensions (5 × 10^4 cells mL^{-1}) stained with Fluo-4 AM (1 μM) and DAPI (3 μM) (λ_{ex} = 490 nm) in the presence of (**a**) thapsigargin (2 μM) or (**b**) EGTA (1.8 mM), recorded at room temperature using a high-pressure cuvette. The applied pressures were sequentially 0.1 MPa (run 1, black curve) → 300 MPa (run 1, blue curve) → 0.1 MPa (run 2, brown curve) → 300 MPa (run 2, purple curve) → 0.1 MPa (run 3, red curve). (**c**) Plots of normalized fluorescence intensities at 518 nm obtained from the data in Fig. 3.3a (black), (**a**) (red), and (**b**) (blue). (Reproduced with permission. Copyright 2021 American Chemical Society)

EGTA did not abolish the depressurization-triggered response and instead produced progressive amplification over repeated cycles (Fig. 3.5b). Comparison of normalized peak intensities across untreated, thapsigargin-treated, and EGTA-treated conditions revealed that suppression of ER refilling attenuated the amplitude somewhat but did not prevent fluorescence jumps, whereas removal of extracellular Ca^{2+} lightly suppressed the depressurization response (Fig. 3.5c). These findings demonstrate that intracellular Ca^{2+} release, likely from ER stores mediated by mechanosensitive intracellular channels or second-messenger cascades, constitutes the dominant source of the observed cytosolic Ca^{2+} transients. While activation of plasma-membrane mechanosensitive channels cannot be entirely excluded, extracellular influx does not appear to be the principal pathway under the tested conditions.

In summary, application of hydrostatic pressure up to 300 MPa to HeLa cells loaded with Fluo-4 AM produced a characteristic response in which depressurization triggered rapid cytosolic Ca^{2+} influxes detectable by enhanced Fluo-4 AM fluorescence, while pressurization itself elicited little immediate fluorescence changes. Pharmacological depletion of ER stores with thapsigargin partially attenuated but did not eliminate depressurization responses, whereas extracellular Ca^{2+} chelation with EGTA failed to prevent the response, identifying intracellular stores as the primary source. The reversible nature of the elevated cytosolic Ca^{2+} after depressurization and the subsequent gradual decay indicate the involvement of cellular homeostatic mechanisms that resequester or remove Ca^{2+}. Collectively, these findings demonstrate that hydrostatic pressure modulation can activate intracellular calcium mobilization in a reversible and measurable manner.

References

Fukuchi, M., Oyama, K., Mizuno, H., Miyagawa, A., Koumoto, K., & Fukuhara, G. (2021). Hydrostatic pressure-regulated cellular calcium responses. *Langmuir, 37*, 820–826.

Hotta, C. T., Gazarini, M. L., Beraldo, F. H., Varotti, F. P., Lopes, C., Markus, R. P., Pozzan, T., & Garcia, C. R. S. (2000). Calcium-dependent modulation by melatonin of the circadian rhythm in malarial parasites. *Nature Cell Biology, 2*, 466–468.

Schneidereit, D., Vass, H., Reischl, B., Allen, R. J., & Friedrich, O. (2016). Calcium sensitive fluorescent dyes Fluo-4 and Fura red under pressure: Behaviour of fluorescence and buffer properties under hydrostatic pressures up to 200 MPa. *PLoS One, 11*, e0164509.

Wei, C., Wang, X., Chen, M., Ouyang, K., Song, L.-S., & Cheng, H. (2009). Calcium flickers steer cell migration. *Nature, 457*, 901–905.

Chapter 4
Pressure-Responsive Molecular Chemosensors

Abstract This chapter provides an integrated overview of how hydrostatic pressure regulates the photophysical pathways and photochemical reactions of chromophores through volumetric changes such as molecular conformation and solvation effects. Across a wide range of systems, pressure modifies not only ground-state properties but also excited-state energy landscapes. Pressurization increases solvent density and viscosity and reorganizes solvation shells, while simultaneously stabilizing compact conformers or promoting intramolecular interactions that are inaccessible at ambient conditions. As a result, pressure can either enhance or suppress fluorescence, invert chiroptical signals, or induce ratiometric fluorescence depending on the topology and flexibility of the scaffold. This chapter also highlights how pressure drives or redirects photochemical pathways. In flexible π frameworks, pressure-induced compaction can gate photochemical reactions by favoring stacked conformers. In multiexcitonic systems, pressurization accelerates singlet fission through a transition-state volume contraction while modulating subsequent triplet dynamics via changes in interactions between solvents and chromophores. These case studies reveal that hydrostatic pressure functions as an orthogonal and predictable thermodynamic parameter that directly couples to molecular volume, offering precise control over emissive states, conformational equilibria, and excited-state reactivity.

Keywords Hydrostatic pressure · Photoreaction · Ratiometric sensor · Chirality · Singlet fission

4.1 Topology-Dependent Inversion of Pressure-Responsive Photoluminescence in Cyclophane Luminophores

Hydrostatic pressure was employed to examine an unconventional, pressure-induced modulation of photoluminescence in a set of closely related molecular luminophores, revealing a striking contrast between cyclic cyclophanes and their acyclic

T. Hamachi, G. Fukuhara, *Hydrostatic Pressure Chemistry*,
https://doi.org/10.1007/978-981-95-7461-2_4

linear analogues. Mechanical stimuli often tune optical responses of organic materials in solution and in condensed phases, and pressure in solution typically enhances fluorescence intensity because solvent viscosity increases and nonradiative collisional quenching is suppressed. Using this common behavior as a benchmark, a chloroform solution of cyclophane, hereafter referred to as compound **1** (Sagara et al., 2017), displayed the opposite response: gradual and substantial fluorescence quenching with increasing hydrostatic pressure up to 320 MPa (Fig. 4.1) (Sagara et al., 2018). In contrast, a linear reference molecule bearing the same 1,6-bis(phenylethynyl)pyrene luminophore, designated compound **2**, exhibited the familiar pressure-dependent fluorescence increase that conforms to viscosity-based models such as the Förster-Hoffmann relationship (Förster & Hoffmann, 1971). Detailed spectroscopic interrogation under controlled high-pressure conditions therefore permits separation of effects that originate from simple viscosity changes from those that arise from structural motifs unique to the cyclophane architecture.

Ambient-pressure characterization established that both compounds absorb in the 350–450 nm region, corresponding to the 1,6-bis(phenylethynyl)pyrene unit, yet subtle distinctions in the ground-state electronic structure emerged from overlaid spectra. The absorption profiles of **1** and **2** were not identical: addition of the naphthalene moiety in cyclophane **1** increased molar extinction between 250 and 330 nm and produced absorption and emission spectra that did not fully overlap with the linear analogue. These spectral differences provide initial evidence for weak intramolecular electronic interactions between the pyrene and naphthalene units in **1** under ambient conditions. Fluorescence spectra recorded under normal conditions displayed a well-resolved vibronic structure, and quantum yield data indicated that the photophysical behavior depends strongly on supramolecular and solid-state organization. In condensed phases, compound **1** adopts distinct emissive states according to thermal history, exhibiting green emission associated with excimer formation in a supercooled nematic liquid crystal and blue emission in the

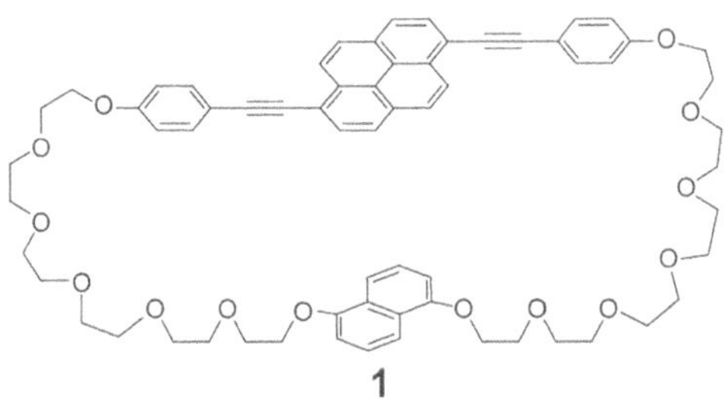

1

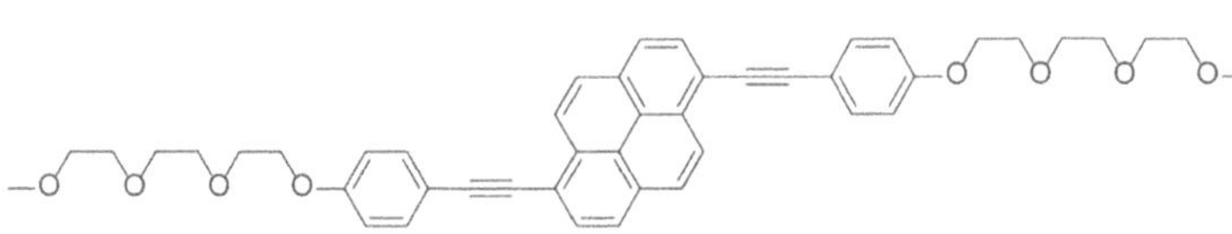

2

Fig. 4.1 Structures of cyclophane **1** and its acyclic reference **2**

crystalline state where excimer formation is suppressed. The solution quantum yield of **1** is high under ambient conditions, underscoring that the observed pressure-induced quenching cannot be attributed to an intrinsically inefficient chromophore.

High-pressure photoluminescence and absorption experiments were performed in the custom-built high pressure cell, which maintains optical access while varying pressure from near atmospheric conditions to several hundred MPa. For chloroform solutions at μM concentrations chosen to avoid aggregation under pressure, compound **1** displayed a steady reduction of emission intensity at the principal 0–0 and 0–1 vibronic bands as pressure increased from 0.1 to 320 MPa (Fig. 4.2a). A modest increase in intensity near 530 nm was observed concurrently, while the principal emission maximum at the 0–0 band bathochromically shifted from 442 to 449 nm. Conversely, the acyclic analogue **2** exhibited the canonical response: emission intensity rose progressively with pressure, with the 0–0 band shifting from 440 to 447 nm and the intensity at 447 nm increasing by a factor of approximately 1.6 at 320 MPa (Fig. 4.2b). The preservation of clear vibronic structures in both systems and the absence of new broad excimer bands support a model in which intermolecular excimer formation between separate luminophores is not the dominant process in these dilute solutions; rather, intramolecular reorganization within cyclophane is implicated.

Pressure-dependent absorption spectra for both compounds show the generally expected red shifts that are common for aromatic chromophores under pressurization (Bovey & Yanari, 1960). Quantitative comparison of spectral slopes, however, highlights the distinct roles of intramolecular interactions in cyclophanes. In chloroform, the slope of the 0–0 absorption wavenumber versus pressure is about $-1.82\ cm^{-1}\ MPa^{-1}$ for the linear compound **2**, whereas cyclophane **1** shows a gentler slope of roughly $-1.20\ cm^{-1}\ MPa^{-1}$. Similar suppression of pressure-induced

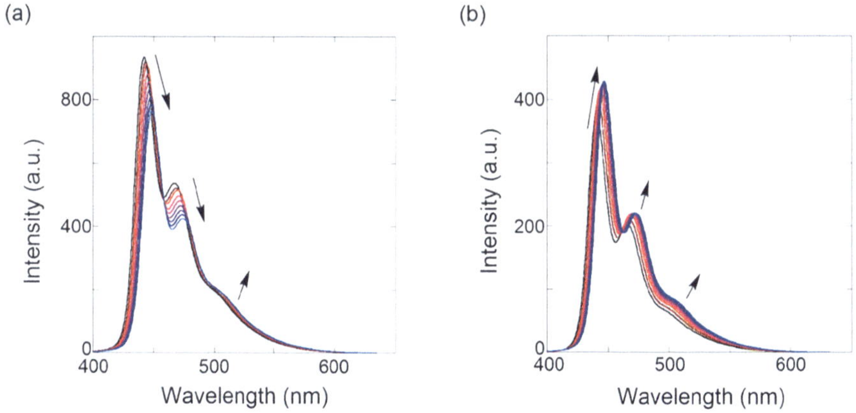

Fig. 4.2 Fluorescence spectra of chloroform solutions of (**a**) **1** (6 μM) and (**b**) **2** (6 μM) recorded at 0.1–320 MPa (black to light blue) at room temperature in a high-pressure cell. Excitation wavelengths were 365 nm for (**a**) and 330 nm for (**b**), and all solutions exhibited nearly identical absorbance intensities. (Reproduced with permission. Copyright 2018 John Wiley & Sons)

bathochromic shifts appears in emission: cyclophane **1** yields slopes near −0.98 and −0.95 cm^{-1} MPa^{-1} for 0–0 and 0–1 emission bands, respectively, while the acyclic **2** exhibits slightly larger slopes of −1.07 and −1.09 cm^{-1} MPa^{-1}. These reduced slopes for **1** are interpreted as evidence that ground-state interactions between the pyrene and naphthalene units become more significant under pressurization, thereby altering how the solvent environment and orbital energies respond to pressure.

Fluorescence intensity plotted against pressure produced distinct trends for the two molecular topologies (Fig. 4.3). For cyclophane **1**, the intensities at 442 and 467 nm decreased linearly as a function of pressure, with correlation coefficients near unity, while the intensity at 530 nm increased slightly (Fig. 4.3a). The linear analogue **2**, on the other hand, showed nonlinear increases in intensity with pressure that collapse onto a single straight line when plotted as log(intensity) versus log(viscosity), thereby conforming to the Förster-Hoffmann expression (Fig. 4.3b, c). This distinction demonstrates that for compound **2**, viscosity-controlled suppression of nonradiative decay dominates, whereas for compound **1**, an additional competing mechanism reduces emission yield as pressure rises.

A mechanistic interpretation consistent with the spectroscopic observations invokes formation of an intramolecular π-stacked conformation in cyclophane **1** that becomes increasingly favored with pressurization. The flexible hexaethylene glycol linkers permit conformational degrees of freedom that allow the naphthalene unit to approach the pyrene chromophore closely, creating an intramolecularly stacked geometry. This geometry supports states of lower fluorescence quantum efficiency and emits at slightly longer wavelengths, accounting for both the quenching of the main vibronic peaks and the modest growth of long-wavelength emission. Compound **2** exhibits the typical pressure-induced red shift of absorption bands, with a slope of −1.82 cm^{-1} MPa^{-1} for the 0–0 absorption transition (Fig. 4.3d, red line). In contrast, compound **1** exhibits a smaller slope of −1.20 cm^{-1} MPa^{-1}, indicating a reduced red shift of the 0–0 absorption band compared with the acyclic reference compound **2** (Fig. 4.3d, black line). This provides clear evidence that significant ground-state interactions between the 1,6-bis(phenylethynyl)pyrene and naphthalene chromophore units are enhanced under pressure, leading to distinct UV-vis spectral behavior and, in turn, opposite fluorescence quenching behavior. Support for the naphthalene-mediated quenching mechanism is provided by a separate quenching experiment in which the free naphthalene chromophore unit was added to solutions of the linear compound **2** at atmospheric pressure; the resulting spectral changes reproduced the quenching pattern, implicating the naphthalene unit and the flexible linker as key contributors.

Generality of the phenomenon extends beyond chloroform. Parallel experiments in solvents spanning a range of polarity, including acetonitrile, toluene, and acetone, revealed a consistent trend: cyclophane **1** consistently showed pressure-induced fluorescence quenching, whereas the linear analogue **2** showed fluorescence enhancement with pressure. Across solvents, cyclophane **1** exhibited systematically gentler pressure-dependent slopes in both absorption and emission relative to the linear molecule, reinforcing the conclusion that intramolecular π interactions

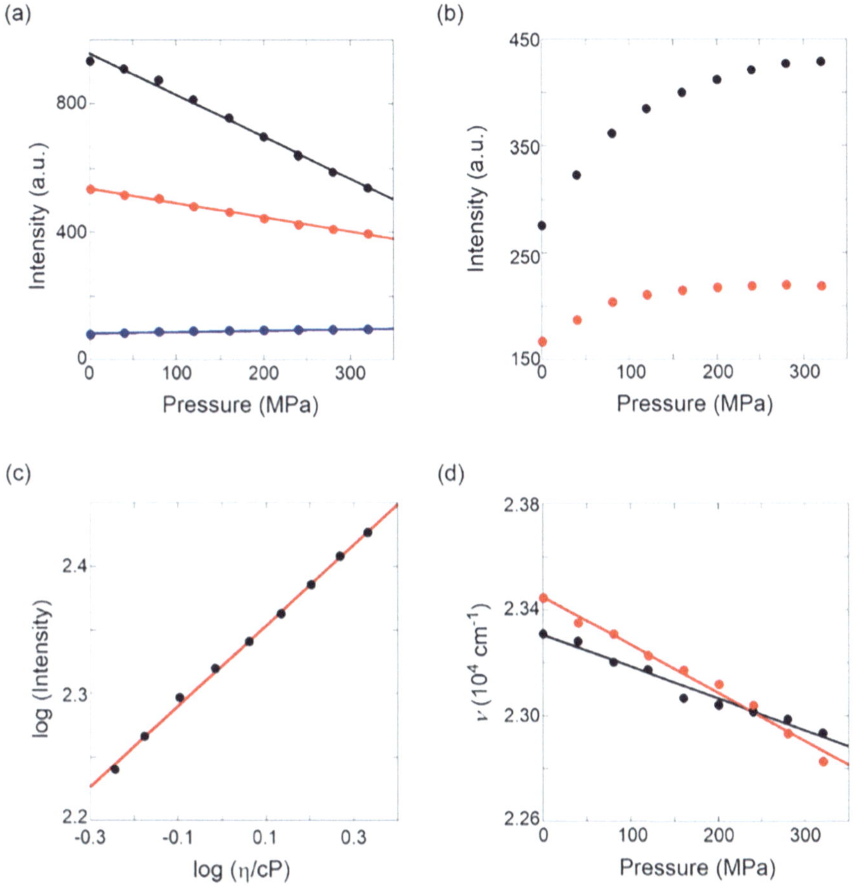

Fig. 4.3 Plots showing the (**a**) pressure-dependent fluorescence intensities of compound **1** in chloroform at 442 nm (black, correlation coefficient $r = 0.996$), 467 nm (red, $r = 0.998$), and 530 nm (blue, $r = 0.941$), and (**b**) those of compound **2** at 447 nm (black) and 473 nm (red). (**c**) Analysis of fluorescence intensity data for compound **2** according to the Förster-Hoffmann equation ($r = 0.999$). (**d**) Pressure dependence of the wavenumber corresponding to the 0–0 absorption maxima of compound **1** (black, $r = 0.986$, slope $= -1.20$ cm^{-1} MPa^{-1}) and compound **2** (red, $r = 0.993$, slope $= -1.82$ cm^{-1} MPa^{-1}) in chloroform. (Reproduced with permission. Copyright 2018 John Wiley & Sons)

between aromatic components within the cyclic scaffold are a robust determinant of pressure response rather than a solvent-specific artifact.

Hydrostatic pressurization therefore biases the conformational free-energy landscape of the flexible cyclophane toward an intramolecularly π-stacked substate of low radiative efficiency. Because the vibronic fine structure remains intact and intermolecular excimer signatures are absent, intramolecular reorganization rather than aggregation best accounts for the pressure-induced fluorescence decrease. Linear analogues that lack a proximate second aromatic unit cannot access this nonemissive intramolecular geometry and therefore exhibit the conventional, viscosity-mediated fluorescence enhancement under pressure. This topology-dependent inversion of pressure response has two practical implications: First, molecular topology must be considered when designing pressure sensors, optical switches, or mechanosensitive probes because the internal degrees of freedom can reverse or modulate photophysical outcomes. Second, measurements on isolated molecules in solution provide complementary mechanistic insight to condensed-phase studies, where grain size, domain heterogeneity, and packing can obscure single-molecule behavior. To translate these mechanistic insights into design principles, future work should combine time-resolved spectroscopy under pressure, computational mapping of conformational energy surfaces as a function of pressurization, and systematic variation of linker length and flexibility. Such approaches will tune the balance between emissive and nonemissive intramolecular geometries, thereby enabling rational development of pressure-tunable luminophores and functional materials, whose optical responses can be inverted or amplified by targeted topological design.

4.2 Reichardt's Dye as a Dual Spectroscopic Indicator of Solvent Polarity and Hydrostatic Pressure

Reichardt's dye (Reichardt & Welton, 2011), a canonical solvatochromic reagent, was evaluated as a dual chemometric indicator capable of reporting both solvent polarity and hydrostatic pressure through measurable changes in its UV-vis absorption spectrum (Miyagawa et al., 2020). Chemical indicators that transduce chemical inputs into spectral, optical, or electrochemical outputs serve as critical tools for probing solvation and related weak interactions. Solvation influences both thermodynamic equilibria and excited-state reaction kinetics, and pressure appears explicitly in thermodynamic expressions through a volumetric term, making hydrostatic pressure a fundamental experimental variable for interrogating solvation effects. Given that Reichardt's empirical E_T scale uses the absorption maximum of a CT dye to quantify solvent polarity (Reichardt, 1994), exploring whether the same chromophore responds sensitively to hydrostatic pressure offered a straightforward route to a dual polarity-and-pressure indicator.

Absorption spectroscopy of dilute solutions of Reichardt's dye (compound **3**) was carried out under controlled hydrostatic pressures up to 300 MPa in eight representative organic solvents spanning the E_T polarity scale from ethyl acetate (38.1 kcal mol^{-1}) to methanol (55.4 kcal mol^{-1}). Under ambient pressure, the spectrum in ethyl acetate featured three principal features: a broad intramolecular CT band denoted **A**, centered broadly in the 500–900 nm region with a full width at half maximum (FWHM) of ~3522 cm^{-1}, a band **B** near 400 nm with FWHM ~3268 cm^{-1}, and a higher energy band **C** around 300 nm with FWHM ~7627 cm^{-1}. Calculation of the empirical E_T parameter from the **A** band reproduced literature values, confirming experimental alignment with established solvatochromic behavior (Fig. 4.4).

Systematic pressurization produced a monotonic hyperchromic increase in absorbance across the spectral window, consistent with an effective increase in dye concentration due to applied pressure (Fig. 4.5). The band widths of **A**, **B**, and **C** remained essentially invariant with pressure, indicating negligible pressure-induced broadening. In contrast, the band maxima displayed markedly different pressure dependences. The low-energy **A** band shifted hypsochromically with increasing pressure, moving from 742 nm at 0.1 MPa to 731 nm at 300 MPa in ethyl acetate. The high-energy **C** band exhibited a small bathochromic shift from 302 to 306 nm over the same pressure range. The **B** band showed negligible peak-position changes. Linear fits of band energy versus pressure yielded well-defined slopes, with the **A**

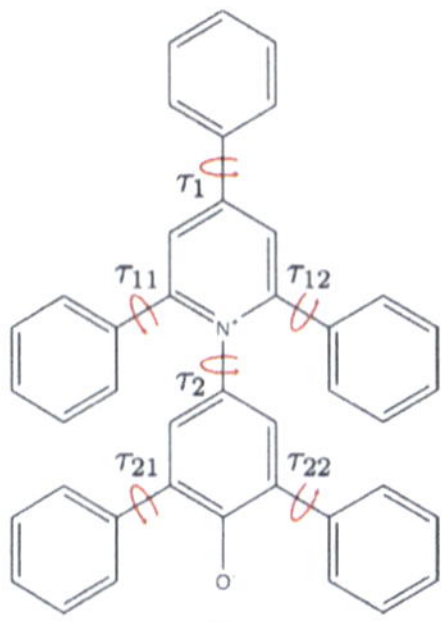

3

Fig. 4.4 Molecular structure of compound **3**. (Reproduced with permission. Copyright 2020 American Chemical Society)

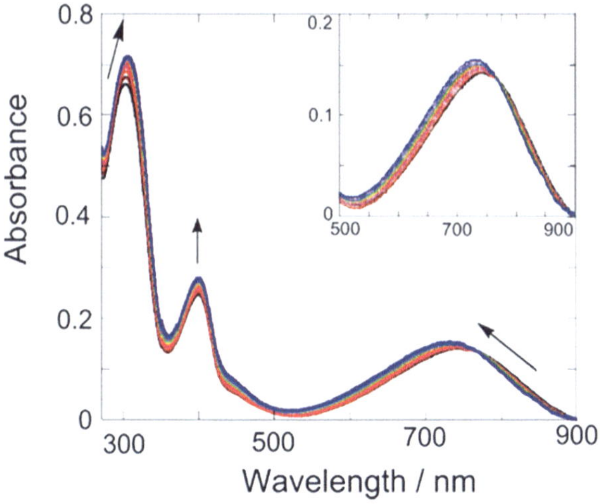

Fig. 4.5 UV-vis absorption spectra of compound **3** in ethyl acetate (158 μM) recorded at room temperature in a high-pressure cell under pressures of 0.1, 20, 40, 60, 80, 100, 120, 140, 160, 180, 200, 220, 240, 260, 280, and 300 MPa. (Reproduced with permission. Copyright 2020 American Chemical Society)

band in ethyl acetate characterized by an α_A of 0.80 cm^{-1} MPa^{-1} and the **C** band by an α_C of −1.40 cm^{-1} MPa^{-1}. Comparable trends were observed across the solvent set, although slopes varied with solvent identity; for example, α_A values in dichloromethane (DCM) and 2-propanol were substantially larger than in ethyl acetate, whereas α_C values showed solvent dependent variability that did not track uniformly with α_A. The empirical E_T parameter calculated at 300 MPa increased relative to ambient pressure values. For instance, Reichardt's E_T in ethyl acetate at 300 MPa approximated the ambient E_T of chloroform, and methanol at 300 MPa registered an E_T comparable to glycol. These observations indicate that hydrostatic pressure can alter the apparent polarity sensed by the dye and that the two spectrally distinct bands can provide complementary references to disentangle pressure and polarity effects.

To elucidate the electronic origins of the experimentally observed spectral responses, time-dependent density functional theory (TD-DFT) using the long-range-corrected functional LRC-ωPBEh was employed, including the lowest 200 singlet states, and molecular dynamics sampling was used to account for ground-state thermal fluctuations. The computed spectrum required an overall bathochromic shift of 134 nm to align with the experiment, after which multiple computed CT transitions clustered to reproduce the experimentally observed broad **A** band. Calculated states at 572, 688, and 930 nm formed this low-energy ensemble. The shorter wavelength bands **B** and **C** were reproduced at approximately 392 and 301 nm and were assigned to mixed local-excitation and CT character for **B**, and principally CT character for **C**. Difference-density plots for representative transitions confirmed that most absorbing states possess significant CT character, with only modest admixture of local excitation in the higher energy bands.

Ab initio molecular dynamics sampling of ground-state conformational space revealed that the equilibrium geometry is quasi C_2 symmetric, with a set of torsional angles τ_i describing rotations of peripheral phenyl and fused-ring groups (Fig. 4.4). Two dominant quasi-normal angle distributions appeared near −47° and +27°, corresponding to two local minima for certain torsions (Fig. 4.6). The variance of the torsions indicated that three angles, corresponding to rotations of bulky, sterically constrained fragments, were relatively rigid with variances around 7°. By contrast, the τ_1 torsion, associated with an end phenyl group, displayed substantially larger flexibility, with variance around 20° (Fig. 4.6a). The overall constrained geometry and the relatively small variances for most torsions argue against a large pressure-driven planarization of the chromophore, a conclusion supported by the experimental observation that FWHM values remain unchanged under pressure. Accordingly, the pressure-induced spectral shifts are unlikely to originate from the marked conformational reorganization of the dye itself.

The differing pressure responses of the **A** and **C** bands were rationalized on the basis of state-specific solvent stabilization linked to dipole-moment differences between the ground and excited states. The ground electronic state possesses a dipole oriented along a particular molecular axis, and increasing solvent polarity stabilizes this state by accommodating its dipole. The principal excited state composing band **A**, namely S_1, corresponds to a CT transition that effectively flips the molecular dipole relative to the ground state. Such a large change in dipole requires

Fig. 4.6 Distributions of the angles (**a**) τ_1 and τ_2, (**b**) τ_{11} and τ_{21}, and (**c**) τ_{12} and τ_{22} obtained from AIMD trajectories in the ground state. (Reproduced with permission. Copyright 2020 American Chemical Society)

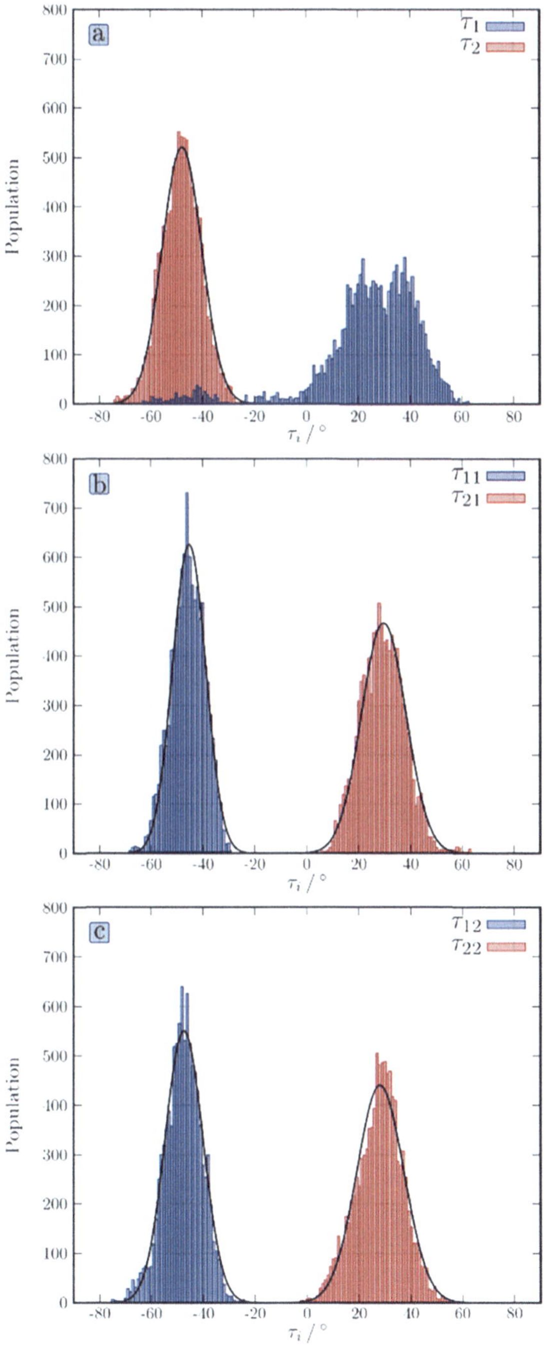

extensive solvent reorganization to fully stabilize the excited state. Under increasing hydrostatic pressure, solvent mobility, and the ability of the solvent molecules to reorient are impaired, reducing the capacity of the medium to reorganize around the excited state dipole. Consequently, the excited state experiences less stabilization relative to the ground state, increasing the transition energy and producing the observed hypsochromic shift of the **A** band. In contrast, the electronic states contributing to the **C** band possess dipole moments that are more similar in direction to the ground state and of lower magnitude. Under pressurization, the ground state is destabilized to a greater extent than these excited states because the solvent environment is displaced from its equilibrium, thus narrowing the energy gap between the ground and excited states associated with the **C** band and yielding a bathochromic shift. Taken together, the state-specific differences in dipole moment direction and magnitude and the pressure dependence of solvent reorganization account for the opposing spectral trends observed for the **A** and **C** bands. Similar mechanistic interpretations have been invoked to describe pressure-induced bathochromic shifts of local excitation-dominated transitions in aromatic systems, suggesting that the interplay between solvent reorganization and electronic-state polarity is a generalizable principle.

In summary, Reichardt's dye exhibits distinct and quantitatively characterizable pressure-induced spectral shifts across multiple solvents, with the long-wavelength CT band shifting to a higher energy under pressurization and the short-wavelength band shifting to a lower energy. This behavior is explained by the differential stabilization of ground and excited states arising from pressure-impaired solvent reorganization and pressure-dependent changes in solvent polarizability. The combination of linear pressure dependences, solvent specific slopes, and supporting TD-DFT and molecular dynamics analyses supports the conclusion that solvent property changes, rather than large molecular conformational changes, dominate the observed spectral modulation, thereby validating the dye as a spectroscopic dual indicator of medium polarity and hydrostatic pressure.

4.3 Ternary Fluorescence Signaling in a Racemic Pseudohelicene Driven by Solvent Polarity and Hydrostatic Pressure

Ternary signaling, in which a single molecular system produces three discrete output states rather than the binary on/off pair, offers a route to higher information density in molecular logic (Ran et al., 2014; He et al., 2015; Zhou et al., 2021), more compact molecular computing elements, and richer sensing modalities that can encode not only the presence or absence but also the directionality of the response. Despite these potential advantages, practical implementations remain limited. Most reported ternary schemes rely on chiral systems that exhibit sign inversion in CD or circularly polarized luminescence (CPL), whereas producing equivalent three-way (positive/zero/negative) outputs from achiral or racemic chemosensors remains challenging. Small organic molecules that reproducibly deliver distinct positive, zero, and negative outputs in response to orthogonal external inputs are rare, and achieving independent control of multiple stimuli within a single molecular platform remains a significant challenge. Hydrostatic pressure represents an attractive stimulus in this context because it is isotropic, orthogonal to common chemical and photophysical inputs, and capable of perturbing both ground- and excited-state equilibria through volumetric effects. A molecular architecture that couples facile conformational switching to modest volume differences between states therefore promises pressure sensitivity together with solvent dependent selectivity. Guided by these considerations, a racemic pseudohelicene scaffold was selected, in which fused anthracene subunits joined by a single bond create a flexible, helical framework whose ground- and excited-state conformational populations are sensitive to solvation and compressive forces (Kinoshita et al., 2025). This design yields a racemic pseudohelicene, **[2]HA$_2$** (Watanabe et al., 2024), that transduces solvent polarity and hydrostatic pressure into positive, zero, and negative fluorescence outputs (Fig. 4.7).

Spectroscopic experiments were conducted in aprotic solvents spanning a polarity range, with toluene, chloroform, and DCM selected as representative media. Solutions were prepared at approximately 60 μM to avoid intermolecular aggregation even under high pressure. Concentration-dependent UV-vis absorption spectra collected between 11 and 179 μM at 0.1, 160, and 320 MPa showed linear absorbance versus concentration and no pressure-induced spectral changes consistent with aggregation or crystallization. Hydrostatic pressure cycles spanning 0.1 to 320 MPa produced fully reversible UV-vis spectra after depressurization, confirming that the system undergoes reversible conformational and solvation changes rather than irreversible phase transitions. A modest pressure-induced bathochromic shift of roughly $-0.7\ cm^{-1}\ MPa^{-1}$ was observed in both absorption and fluorescence

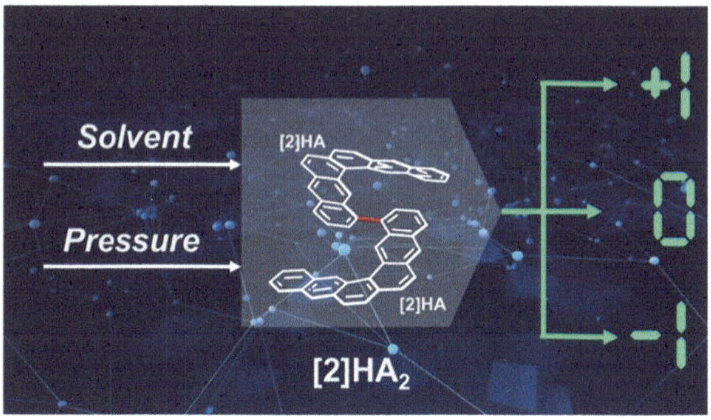

Fig. 4.7 Schematic of the ternary signal transduction in **[2]HA$_2$** triggered by hydrostatic pressure and solvent as multidimensional stimuli. (Reproduced with permission. Copyright 2025 Royal Society of Chemistry)

band positions, a magnitude comparable to common π-conjugated dyes and therefore insufficient by itself to explain the unique signaling behavior.

Fluorescence intensity responded to hydrostatic pressure in a strongly solvent-dependent manner. A pressure-induced intensity change, defined as $\Delta I = I_{280\,\text{MPa}} - I_{0.1\,\text{MPa}}$ (using the maximum at 280 MPa), was +228 in toluene, +92 in chloroform, and −235 in DCM, yielding a clear inversion of response sign as solvent polarity increased. Solvent polarity was parameterized by Reichardt's empirical $E_T(30)$ scale, with values of 33.9, 39.1, and 40.7 kcal mol^{-1} for toluene, chloroform, and DCM, respectively. Plotting ΔI against $E_T(30)$ revealed a polarity-dependent crossover near $E_T(30)$ ~40 kcal mol^{-1}, indicating a transition from a positive to a negative pressure response as solvent polarity increases (Fig. 4.8).

Fluorescence lifetime and absolute quantum yield measurements under pressure provided insight into the competing radiative and nonradiative processes that underpin the intensity changes. Absolute fluorescence quantum yields at ambient pressure were comparable across solvents but evolved differently under pressurization. Fluorescence lifetimes increased from 9.1 to 11.7 ns in toluene between 0.1 and 320 MPa, showed a minor increase in chloroform, and remained essentially unchanged in DCM. From the measured quantum yields and lifetimes, radiative (k_F) and nonradiative (k_{nr}) rate constants were calculated, and their pressure sensitivities were quantified by fitting ln k_F and ln k_{nr} versus pressure to obtain slopes denoted α and β, respectively (Fig. 4.9). In toluene, α was slightly positive (+1.1 × 10^{-4} MPa^{-1}) while β was strongly negative (−13.5 × 10^{-4} MPa^{-1}), indicating that inhibition of nonradiative decay dominates under pressurization and thus enhances emission. In DCM, the trends were reversed, with α = −11.2 × 10^{-4} MPa^{-1} and β = +3.8 × 10^{-4} MPa^{-1}, consistent with pressure-activated nonradiative pathways that quench fluorescence. Chloroform displayed intermediate behavior, with both slopes of comparable magnitude and modest net effect on emission. These quantitative kinetic signatures establish that pressure can either suppress or promote radiationless decay depending on solvent-mediated conformational preferences.

DFT calculations at the CAM-B3LYP/def2-SVP level with SMD solvent corrections identified three low energy diastereomeric conformers of **[2]HA$_2$** arising from combinations of helical handedness for each fused anthracene subunit and axial chirality around the connecting single bond. Conformers are denoted here by triple labels such as (P, P_a, P), (M, P_a, P), and (M, P_a, M) (Fig. 4.10a). Barrier calculations indicate a large activation free energy for axial chirality inversion, on the order of 102.4 kJ mol^{-1}, whereas interconversion among the three low energy conformers proceeds over substantially lower barriers, with a maximum of 16.8 kJ mol^{-1}. These energetic features suggest kinetically persistent axial chirality together with facile conformer exchange involving the helical subunits. Relative energies and molecular volumes for each conformer were computed in both the ground and first excited states, revealing solvent-dependent stability patterns and modest volume differences on the order of a few Å^3. In the ground state at ambient pressure, (M, P_a, M) is most stable in toluene while (P, P_a, P) is most stable in DCM, and rapid

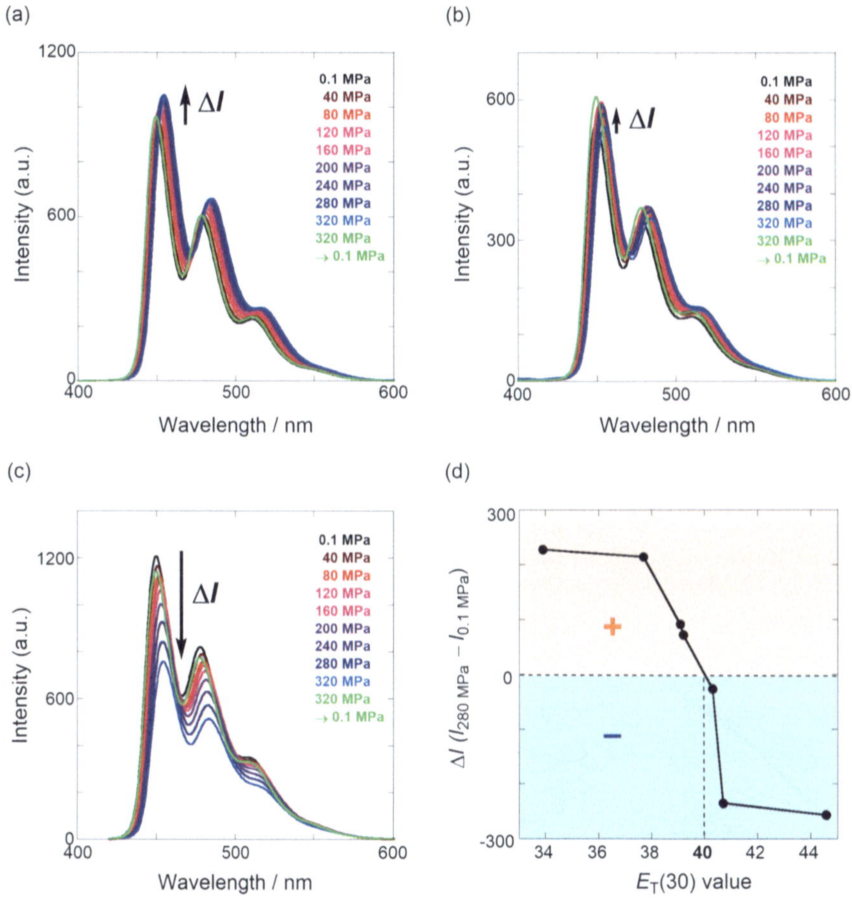

Fig. 4.8 Fluorescence spectra of **[2]HA$_2$** recorded under varying pressures in (**a**) toluene (λ_{ex} = 391 nm, 66 μM), (**b**) chloroform (λ_{ex} = 390 nm, 61 μM), and (**c**) DCM (λ_{ex} = 412 nm, 57 μM) at room temperature using a high-pressure cell. Spectra were obtained at 0.1, 40, 80, 120, 160, 200, 240, 280, and 320 MPa (black to light blue). The spectrum at 0.1 MPa after release from 320 MPa is shown in green. (**d**) Pressure-induced changes in fluorescence intensity, expressed as ΔI ($I_{280\ \mathrm{MPa}} - I_{0.1\ \mathrm{MPa}}$), plotted against $E_T(30)$. (Reproduced with permission. Copyright 2025 Royal Society of Chemistry)

conformational interconversion yields equilibrium populations that reflect these solvent-dependent energy biases.

Combining experimental kinetics and calculated energetics yields a coherent mechanistic model for the ternary signaling. Excitation from a ground-state conformer into the Franck-Condon manifold allows relaxation to different excited-state conformers, thereby selecting emissive or radiationless decay pathways whose efficiencies depend on conformer identity. In toluene at ambient pressure, excitation of the (*M*, P_a, *M*) ground state favors relaxation through an excited-state pathway that converts to the (*P*, P_a, *P*) excited conformer, a pathway associated with enhanced

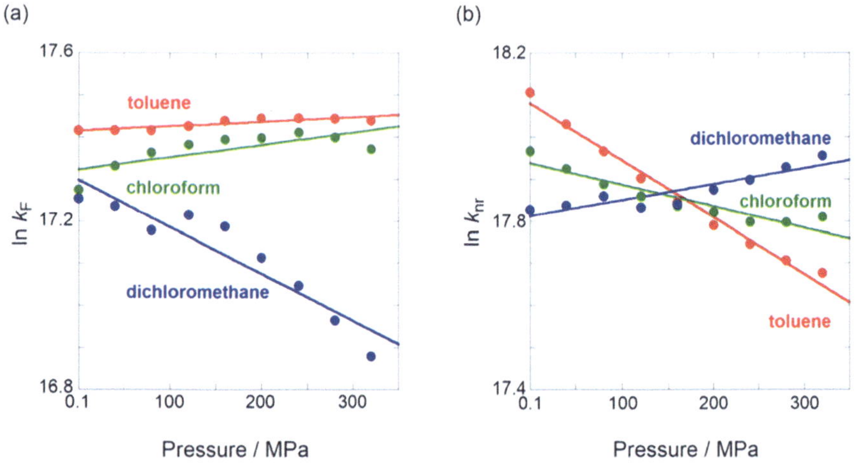

Fig. 4.9 Pressure dependence of (**a**) k_F in toluene (red, $r = 0.878$), chloroform (green, $r = 0.747$), and DCM (blue, $r = 0.937$), and (**b**) k_{nr} in toluene (red, $r = 0.992$), chloroform (green, $r = 0.938$), and DCM (blue, $r = 0.905$), all measured at room temperature. (Reproduced with permission. Copyright 2025 Royal Society of Chemistry)

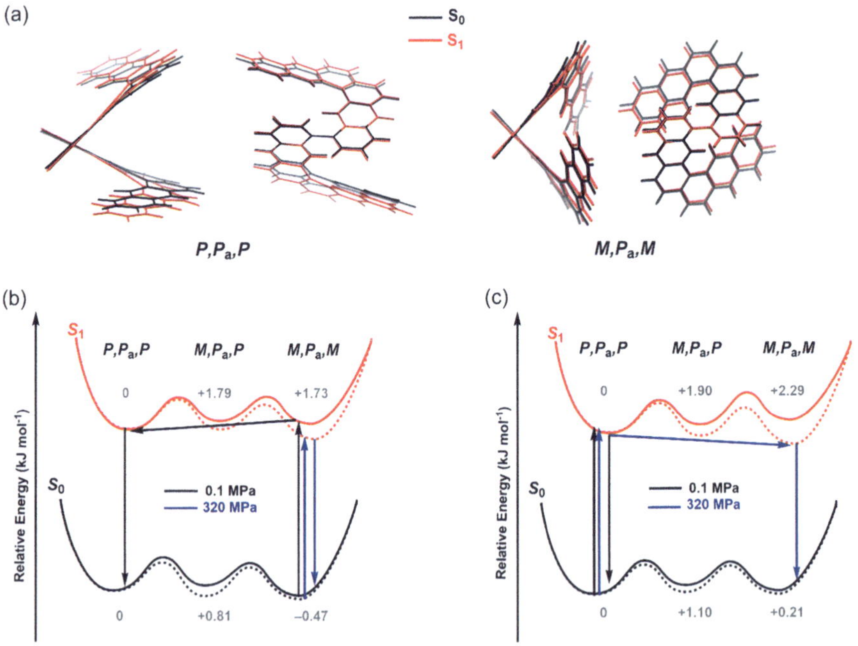

Fig. 4.10 (**a**) Optimized structures of **[2]HA$_2$** in the S_0 (black) and S_1 (red) states in DCM. (**b**, **c**) Schematic energy diagrams of **[2]HA$_2$** in (**b**) toluene and (**c**) DCM. Dotted lines indicate hypothetical energy profiles under high-pressure conditions. Arrows indicate the hypothetical major components contributing to the emission process. (Reproduced with permission. Copyright 2025 Royal Society of Chemistry)

nonradiative deactivation and correspondingly lower fluorescence quantum yield (Fig. 4.10b). Under hydrostatic pressure, volume reduction disfavors a conformational change that increases excited-state volume, thereby suppressing conversion to the nonemissive pathway and increasing the fluorescence quantum yield. This pressure-induced suppression manifests as a positive ΔI response. In contrast, in DCM at ambient pressure, emission primarily arises from the (P, P_a, P) conformer, which exhibits relatively efficient radiative decay (Fig. 4.10c). Pressurization alters the excited-state landscape so that relaxation toward the more compact (M, P_a, M) conformer is promoted, increasing nonradiative decay and reducing fluorescence intensity, thereby producing a negative ΔI response. Chloroform occupies an intermediate regime in which solvent stabilization and pressure-induced volumetric effects balance, yielding little net intensity change under pressurization.

The discovery of ternary signaling in the racemic pseudohelicene **[2]HA$_2$** expands the toolbox for molecule-based logic and sensing by demonstrating that small, achiral organic molecules can deliver positive, zero, and negative outputs in response to orthogonal stimuli. Whereas most prior molecular ternary systems have relied on chiral observables such as CD or CPL sign inversion, or on large supramolecular and macromolecular scaffolds that complicate synthesis and integration, the present system shows that careful design of flexible helical subunits connected by a single bond, together with solvent-dependent conformational energetics and modest volumetric differences between conformers, can generate multidimensional, reversible signaling. By elucidating a mechanistic basis grounded in excited-state conformational dynamics, this study establishes a foundation for developing piezosensitive molecular devices, molecular computing elements, and pressure-responsive probes.

4.4 Pressure-Induced Intramolecular Photodimerization in Helical Anthracenes for Ratiometric Chemosensing

This section presents helicene derivatives based on a helical anthracene scaffold as candidates for pressure-responsive chemosensors. Anthracene is known to undergo [4 + 4] photocyclodimerization to afford products with distinct spectral properties (Jezowski et al., 2012), thereby enabling ratiometric sensing through the fluorescence intensity ratio between the monomer and its dimerized species. A helical framework such as helicene, which can undergo expansion and contraction motions analogous to a molecular spring, was employed as the structural platform for anthracene, as it is expected to be highly susceptible to conformational changes under hydrostatic pressure (Kinoshita et al., 2022). A series of anthracene-fused helicenes, including **[3]HA**, **[3]HA-Ph**, **[3]HA2NAP**, **[4]HA**, and **[5]HA** (Fig. 4.11) were prepared, and their photophysical properties were systematically investigated under hydrostatic pressure to evaluate the feasibility of detecting ratiometric fluorescence changes.

Initial hydrostatic pressure-dependent UV-vis, fluorescence, and excitation measurements revealed that all HAs exhibited modest bathochromic shifts under

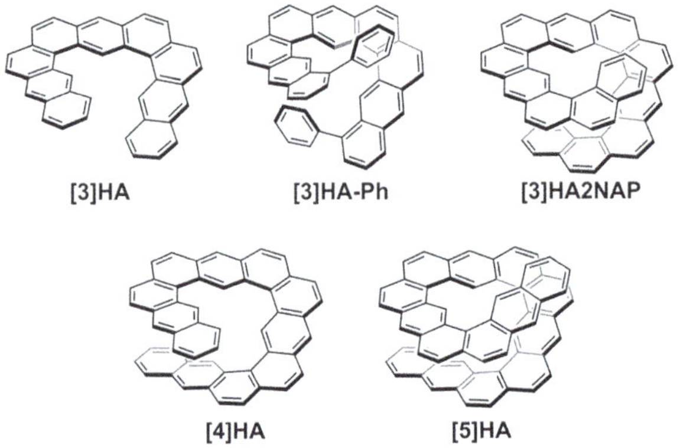

Fig. 4.11 Molecular structures of helical anthracenes (**[3]HA**, **[3]HA-Ph**, **[3]HA2NAP**, **[4]HA**, and **[5]HA**). (Reproduced with permission. Copyright 2022 Royal Society of Chemistry)

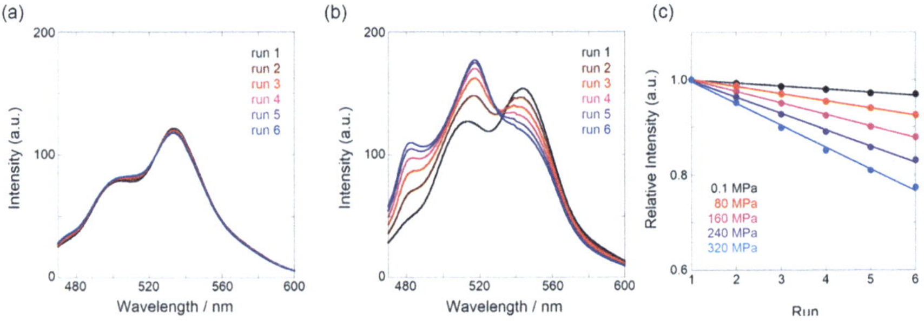

Fig. 4.12 Fluorescence spectra of toluene solutions of 95 μM **[4]HA** (λ_{ex} = 456 nm) recorded at (**a**) 0.1 MPa and (**b**) 320 MPa at room temperature in a high-pressure cell. Spectra were obtained upon photoexcitation for the first (black), second (brown), third (red), fourth (pink), fifth (purple), and sixth (blue) irradiation runs. (**c**) Relative fluorescence intensity ratios ($I_{run\ x}/I_{run\ 1}$) of **[4]HA** at 0.1, 80, 160, 240, and 320 MPa. (Reproduced with permission. Copyright 2022 Royal Society of Chemistry)

pressure, consistent with the stabilization of π* orbitals. However, only **[4]HA** displayed an unexpected pressure-dependent photoreactivity that produced clear ratiometric fluorescence changes. At atmospheric pressure (0.1 MPa), the fluorescence spectrum of **[4]HA** was essentially unchanged over six repeated photoexcitation runs (Fig. 4.12a). In contrast, at 320 MPa, the spectrum evolved markedly with repeated photoexcitation, showing a progressive decrease in intensity near 545 nm and concomitant increases near 515 and 480 nm (Fig. 4.12b). At intermediate pressures of 80, 160, and 240 MPa, the same type of spectral change was observed, but the rate of change depended on the applied pressure. The relative fluorescence intensity ratio at 545 nm for each run was plotted against the run number, showing linear decreases up to the sixth run for each pressure (Fig. 4.12c). The slopes of those plots increased monotonically with pressure. Because the ratiometric fluorescence intensities plotted against each run produced linear relationships, the hydrostatic pressure in the targeted medium can be quantitatively determined.

To elucidate whether ratiometric photoreactivity is a general property of anthracene-fused helicenes or a special feature of **[4]HA**, the same repeated photoexcitation experiments at 320 MPa were performed on **[3]HA**, **[3]HA2NAP**, and **[5]HA**. None of these reference compounds showed the progressive ratiometric spectral changes observed for **[4]HA**. This selectivity suggests that the unique structural characteristics of **[4]HA** are responsible for the observed behavior, in particular the presence of an intramolecular overlap between the fused anthracene termini. Under hydrostatic pressurization, this overlap is promoted, leading to the formation of a stacked intramolecular conformer in which the π systems of the terminal anthracenes are closely aligned. Upon photoexcitation, this stacked conformer undergoes an intramolecular [4 + 4] photocyclodimerization reaction, resulting in a covalently cross-linked dimer. The formation of the dimer substantially alters both the electronic absorption spectrum and the fluorescence emission profile of the molecule, giving rise to the distinct ratiometric response observed under high-pressure conditions.

To test this mechanistic picture, computational and experimental spectroscopic analyses were performed. DFT calculations at the B3LYP/6-31G(d,p) level were used to generate optimized structures for the monomeric **[4]HA** and the proposed intramolecular photodimer **[4]HA-C**, in which the anthracene units are cross-linked at the 9 and 10 positions (Fig. 4.13a). TD-DFT-derived oscillator strengths produced simulated UV-vis spectra with an absorption maximum predicted at 383 nm for **[4]HA** (Fig. 4.13b) and at 363 nm for **[4]HA-C** (Fig. 4.13c), corresponding to a theoretical hypsochromic shift of approximately 20 nm upon formation of the dimer. Experimentally, excitation spectra measured after performing ten cycles of photoexcitation at 320 MPa and monitoring different emission bands revealed a new excitation peak near 352 nm when monitoring the 482 nm emission, whereas the original monomer band remained near 370 nm. This experimentally observed hypsochromic shift of 18 nm matches closely with the theoretical prediction, supporting the assignment of the new shorter wavelength excitation and the associated 482 nm emission to the intramolecular photodimer **[4]HA-C**. Thus, the ratiometric evolution of fluorescence is attributed to the stepwise conversion of monomeric **[4]HA** into the intramolecular photodimer, with the conversion rate enhanced by increasing hydrostatic pressure.

Fluorescence lifetimes were measured under pressure to provide further mechanistic insight. For HAs that lack overlapping anthracene termini, such as **[3]HA** and **[3]HA-Ph**, a single excited-state lifetime of roughly 6–9 ns was observed, which increased under pressure. This pressure-induced lifetime elongation is consistent with the suppression of collisional deactivation by the solvent as compressive conditions alter solvation dynamics. For **[3]HA2NAP** and **[4]HA**, which possess structural overlap of the anthracene termini, multi-exponential decay behavior was observed, consisting of a major longer-lived component in the 9–14 ns range and a minor short-lived component in the sub-ns to a few ns range. Importantly, the relative abundance of the short-lived component increased upon pressurization. For example, it rose from 0.02 to 0.20 for **[3]HA2NAP** and from 0.04 to 0.08 for **[4]HA** upon increasing pressure to 320 MPa. Multi-wavelength lifetime analysis and

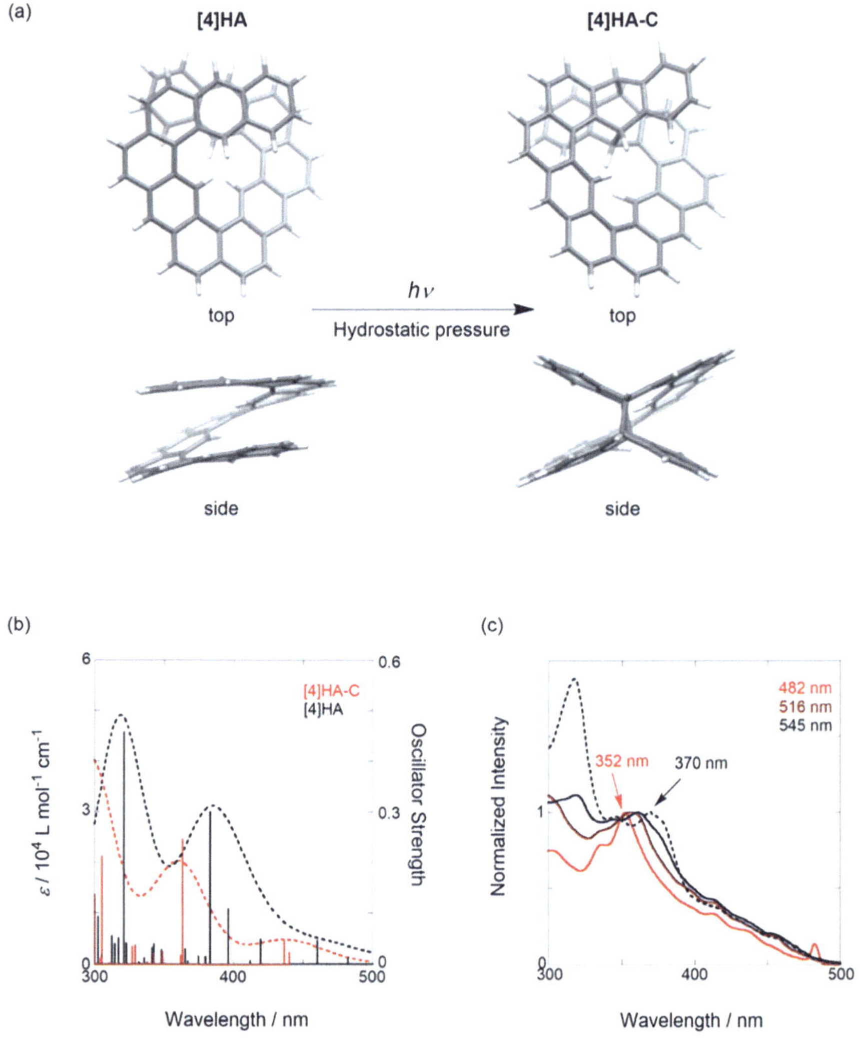

Fig. 4.13 (**a**) X-ray crystal structure of **[4]HA** (left) and the geometry-optimized structure of **[4]HA-C** (right) obtained using the B3LYP/6-31G(d,p) method. (**b**) Calculated oscillator strengths (bars) and simulated UV-vis absorption spectra (dotted lines) for **[4]HA** (black) and **[4]HA-C** (red), with a half-width of 0.20 eV applied for spectral broadening. (**c**) Normalized excitation spectra of 95 μM **[4]HA** in toluene at 320 MPa, monitored at 545 nm (black), 516 nm (brown), and 482 nm (red). The black dotted line shows the normalized UV-vis absorption spectrum of **[4]HA** at 320 MPa. (Reproduced with permission. Copyright 2022 Royal Society of Chemistry)

time-resolved emission spectra showed that the short-lived component emits at shorter wavelength than the longer-lived species. Combining these experimental observations with the theoretical assignments, the short-lived, blue-shifted emitting species is interpreted as an intramolecularly stacked conformer of the fused anthracene ends, a precursory intermediate that rapidly undergoes intramolecular photocyclodimerization. The formation of this stacked intermediate is favored by hydrostatic pressurization, explaining why the photoreaction and consequent ratiometric fluorescence changes accelerate with pressure.

[5]HA also presents overlapping anthracene termini but did not show evidence of the stacking-related short-lived excited species or pressure-induced ratiometric photochemistry. This indicates that the formation of a stacked conformer suitable for intramolecular [4 + 4] photocyclodimerization is hindered in **[5]HA** due to steric repulsion within its larger fused anthracene framework. This contrasting behavior highlights that helical pitch and the number of fused anthracene units are crucial structural parameters that determine whether pressure-induced intramolecular stacking and photodimerization will occur.

In conclusion, the present study identifies a new mechanism for pressure-responsive ratiometric chemosensing based on an intramolecular π-stacked conformer and its subsequent photoreaction in the helical anthracene moiety. Hydrostatic pressure facilitates the formation of an intramolecular π-stacked conformer, which then undergoes rapid [4 + 4] photocyclodimerization to yield a blue-shifted and distinct fluorescent species. The ratiometric fluorescence exhibits a clear dependence on pressure, and the relative fluorescence intensity changes show a linear relationship with the photoexcitation run number at each pressure level. Collectively, these features provide a straightforward calibration method for estimating unknown pressures in solution. These findings show that helical pitch and π-extension can be tuned to control pressure-responsive conformational changes and photoreactivity. Therefore, the rational design of π-extended molecules could drive significant progress in the development of pressure-sensitive chemosensors.

4.5 Pressure-Driven Control of Excited-State Conformational Equilibria in the Donor-Acceptor-Donor Chromophore

Understanding how external stimuli, such as hydrostatic pressure, influence the excited-state properties of luminophores represents a crucial step in the development of functional optical materials. Such systems are increasingly important in the design of sensors, imaging probes, and mechanochromic devices, where emission color or intensity can be modulated in response to environmental changes. A particularly promising candidate is **PTZ-DBPHZ** (Okazaki et al., 2017), a donor-acceptor-donor (D-A-D) compound containing two phenothiazine (PTZ) moieties bound to a dibenzo[*a,j*]phenazine (DBPHZ) core (Fig. 4.14). This molecule exhibits conformational diversity, adopting several ground-state geometries that determine its photophysical behavior. Because the relative populations of these conformers and their excited-state dynamics can be perturbed by external influences, **PTZ-DBPHZ** provides an exemplary model to explore the ratiometric luminescence responses induced by pressure (Takeda et al., 2019).

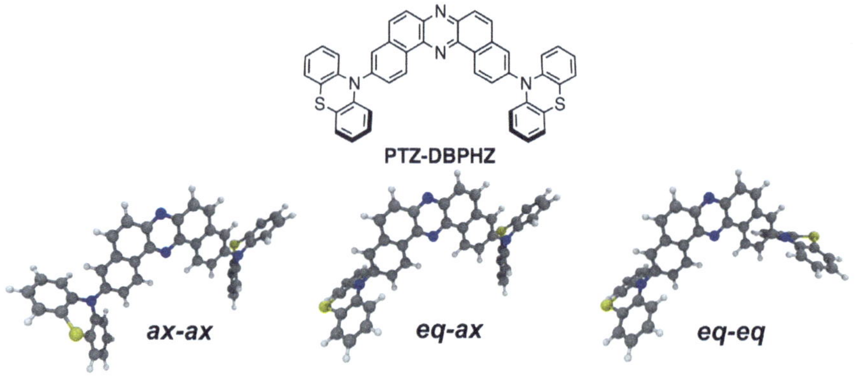

Fig. 4.14 Molecular structure of **PTZ-DBPHZ** and different conformers (quasi axial-quasi axial *ax-ax*; quasi equatorial-quasi axial *eq-ax*; quasi equatorial-quasi equatorial *eq-eq*). (Reproduced with permission. Copyright 2019 John Wiley & Sons)

In general, ratiometric luminophores have attracted significant attention owing to their ability to simultaneously generate emissions from two or more distinct excited states. In typical donor-acceptor dyads, excitation leads to a locally excited (LE) emission at a shorter wavelength and a CT emission at a longer wavelength, with the relative intensities governed by solvent polarity or viscosity. Structural reorganization in the excited state, particularly the twisting between donor and acceptor groups, produces the TICT state that defines their response properties. Another important class includes bent π-conjugated systems such as *N*,*N*′-disubstituted dihydrophenazines and cyclooctatetraene-fused acenes, which undergo planarization upon excitation. In such compounds, the degree of planarization is highly environment-dependent, resulting in tunable emission.

PTZ-DBPHZ offers a conformationally dictated regulation mechanism. At least four conformers, including axial-axial (*ax-ax*, both *syn* and *anti*), equatorial-axial (*eq-ax*), and equatorial-equatorial (*eq-eq*), can coexist in solution. Previous time-resolved studies established that the conformational distribution governs the photophysics in the solid state, motivating investigation into whether external stimuli such as hydrostatic pressure can control the equilibria among excited-state conformers in solution. Pressure is especially well-suited as an external stimulus because it alters solvent density, refractive index, and viscosity, thereby tuning solvation dynamics and relaxation pathways without changing the chemical composition.

Spectroscopic measurements of **PTZ-DBPHZ** in toluene and methylcyclohexane (MCH) under pressures up to 350 MPa reveal a striking ternary emission behavior comprising LE, CT1, and CT2 components. In toluene at ambient pressure, three peaks are observed at 481, 508, and 650 nm, respectively, assigned to LE emission of *ax-ax* conformers and CT emissions associated with *eq-ax* and *eq-eq* conformers (Fig. 4.15a). TD-DFT calculations support these assignments, correlating the observed spectra with predicted excited-state energies. Notably, the excitation wavelength significantly influences the emission profile. Excitation at 430 nm yields comparable intensities from all three states, whereas higher-energy excitation at 400 nm selectively enhances CT contributions. This behavior is consistent with excitation energies enabling conformational rearrangement in the excited state.

The effect of increasing hydrostatic pressure on the emission behavior in toluene is notable. As pressure rises, the CT2 emission is progressively suppressed, while the intensities of the LE and CT1 emissions increase (Fig. 4.15b, d). Peak positions also shift bathochromically at rates near -1 cm^{-1} MPa^{-1}, reflecting pressure-induced increases in solvent polarizability and dielectric constant (Fig. 4.15c). Despite these excited-state changes, the absorption and excitation spectra remain relatively stable, implying that ground-state conformer populations are largely unaffected. Thus, pressure modulates the equilibria only in the excited-state landscape. This distinction emphasizes that the observed ratiometric luminescence is a result of pressure-dependent relaxation kinetics rather than ground-state conformational redistribution.

Kinetic modeling provides mechanistic insight into these behaviors. The equilibrium among the LE, CT1, and CT2 states can be described by competing forward and backward rate constants. Analysis of the pressure dependence of quantum yield ratios reveals that CT1 emission undergoes a regime change around 200 MPa: at lower pressures, the Φ_{CT1}/Φ_{LE} ratio increases with pressure, indicative of thermodynamic control where the back reaction dominates, while at higher pressures it decreases, signifying kinetic control where the forward reaction becomes rate-limiting. In contrast, the CT2 emission relative to LE follows a single linear trend across the pressure range, pointing to persistent kinetic control. The observed activation volume differences ($\Delta V^{\neq}_{obs}$) were positive for both the CT1-LE and CT2-LE processes, indicating that the aromatic toluene molecules effectively solvate the **PTZ-DBPHZ** chromophore. Under elevated pressure, such enhanced solvation is expected to promote the formation of transient solvent clusters around the luminophore. The dynamic solvent effect arising from this clustering can partially restrict the conformational rearrangement of the donor units, thereby contributing to the positive activation volumes. Notably, the $\Delta V^{\neq}_{obs}$ value was significantly larger for the CT2-LE conversion (9.3 cm^3 mol^{-1}) than for the CT1-LE process (2.9 cm^3 mol^{-1}), suggesting preferential solvation of the *eq-eq*-type conformer associated with the CT2 state. This interpretations is fully consistent with the pronounced decrease in CT2 emission observed upon pressurization (Fig. 4.16).

The viscosity dependence is further corroborated by the Förster-Hoffmann equation (Fig. 4.17). The LE and CT1 emission intensities both scale with solvent viscosity; however, CT1 exhibits a sensitivity factor more than twice that of LE, consistent with the more polar nature of the CT state. This enhanced sensitivity highlights the potential of **PTZ-DBPHZ** as a viscosity probe analogous to established molecular rotors, while offering additional ratiometric advantages.

Measurements in MCH provide a complementary perspective. Because MCH is less polar than toluene, it alters conformer energetics by preferentially stabilizing *eq-eq* conformers with smaller dipole moments relative to *eq-ax*. The absorption spectrum in MCH was similar to that in toluene, and no significant changes were

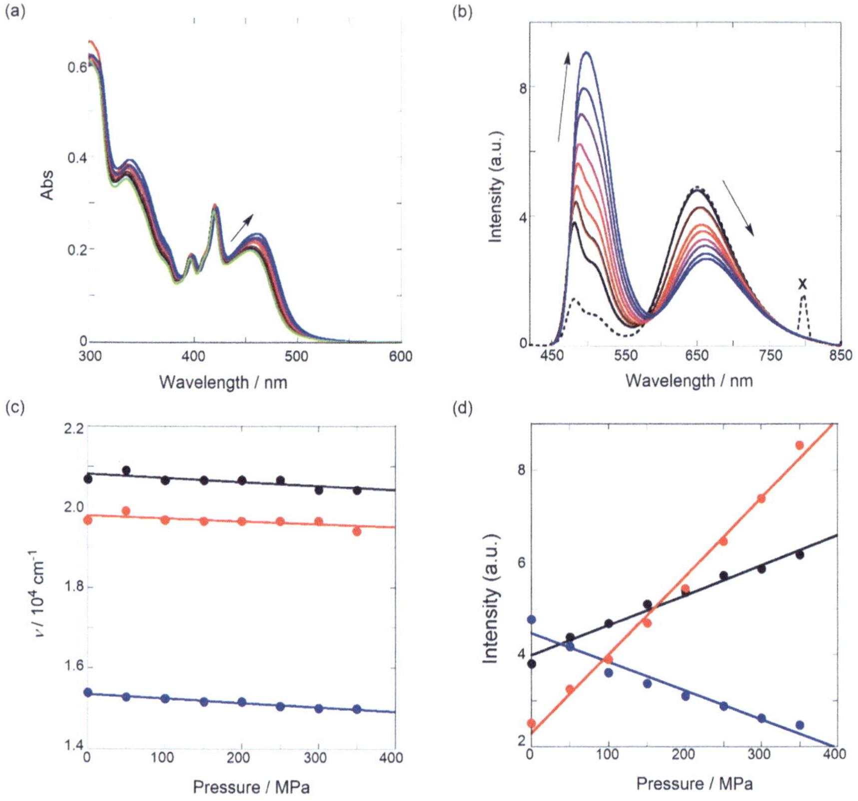

Fig. 4.15 (**a**) UV-vis absorption and (**b**) fluorescence spectra (excited at 430 nm) of **PTZ-DBPHZ** (222 μM) in toluene at room temperature under pressures of 0.1, 50, 100, 150, 200, 250, 300, and 350 MPa (black to blue lines). The dotted line indicates the spectrum at 0.1 MPa excited at 400 nm, and the light-green line represents the spectrum after depressurization from 350 MPa. All measurements were performed using a high-pressure optical cell. (**c**) Plots of fluorescence maxima for the LE band (black; r = 0.797, slope = -1.05 cm^{-1} MPa^{-1}), CT1 band (red; r = 0.691, slope = -0.748 cm^{-1} MPa^{-1}), and CT2 band (blue; r = 0.984, slope = -1.10 cm^{-1} MPa^{-1}). (**d**) Pressure dependence of fluorescence intensities at 481 nm (black; r = 0.990, slope = 0.00653 MPa^{-1}), 508 nm (red; r = 0.996, slope = 0.0170 MPa^{-1}), and 645 nm (blue; r = 0.974, slope = -0.00622 MPa^{-1}). (Reproduced with permission. Copyright 2019 John Wiley & Sons)

observed upon pressurization, indicating that the conformational changes in the ground state are minimal (Fig. 4.18a). At ambient pressure, the emission spectrum in MCH is dominated by CT2 at 571 nm (*eq-eq* conformer). Calculated conformer populations confirm the solvent effect, predicting a higher fraction of *eq-eq* compared to *eq-ax*. Thus, solvent polarity dictates the balance between CT1 and CT2 emissions.

The pressure response in MCH differs markedly from that in toluene. Rather than ratiometric shifts, the intensities of all emission bands increase with pressure, while peak positions shift modestly (Fig. 4.18b–d). This behavior suggests that in

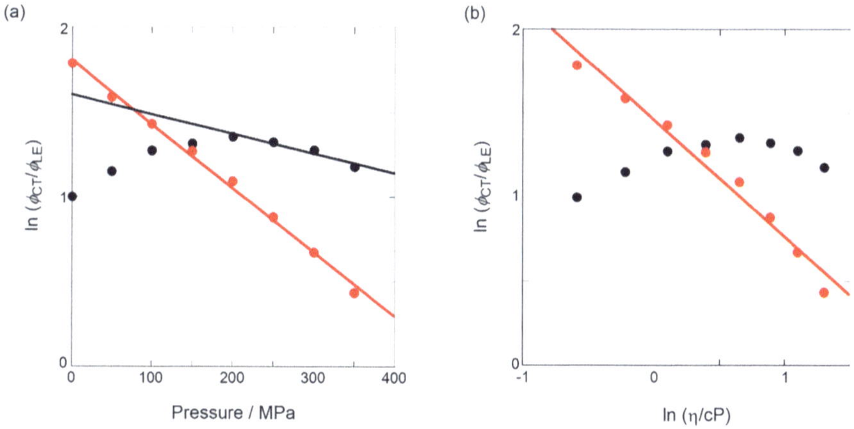

Fig. 4.16 Plots of the natural logarithm of the quantum yield ratios (black, CT1/LE; red, CT2/LE) as a function of (**a**) pressure (black: $r = 0.963$, slope = -0.00116 MPa^{-1} up to 200 MPa; red: $r = 0.998$, slope = -0.00377 MPa^{-1}) and (**b**) ln(η/cP) (red: $r = 0.987$, slope = -0.69). (Reproduced with permission. Copyright 2019 John Wiley & Sons)

Fig. 4.17 Correlation plots of fluorescence intensities at 481 nm (black; LE, $r = 0.994$, $B = 0.248$, $C = 0.656$) and 508 nm (red; CT1, $r = 0.999$, $B = 0.636$, $C = 0.565$) for **PTZ-DBPHZ** in toluene, fitted using the Förster-Hoffmann equation ([$\log(I) = B \log(\eta) + C$]). (Reproduced with permission. Copyright 2019 John Wiley & Sons)

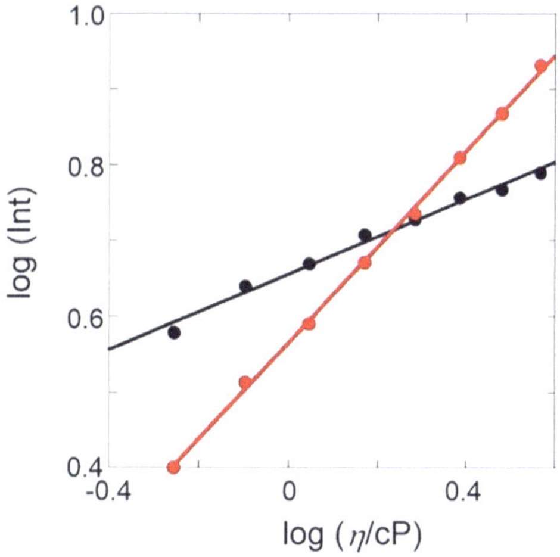

MCH, the major effect of pressure is the suppression of nonradiative decay, likely by restricting molecular vibrations, rather than rebalancing conformer-dependent excited-state pathways. Consequently, while toluene highlights the ability of pressure to tune excited-state equilibria, MCH emphasizes its role in modulating nonradiative deactivation. Both cases illustrate complementary facets of how hydrostatic pressure governs luminescence.

To further clarify the influence of hydrostatic pressure on the excited-state dynamics of **PTZ-DBPHZ**, time-correlated fluorescence decays were recorded in toluene and MCH under pressurized conditions. In toluene, the decay at 490 nm could be deconvoluted into two components: a short-lived species (τ_2 ~ 1.3–1.5 ns)

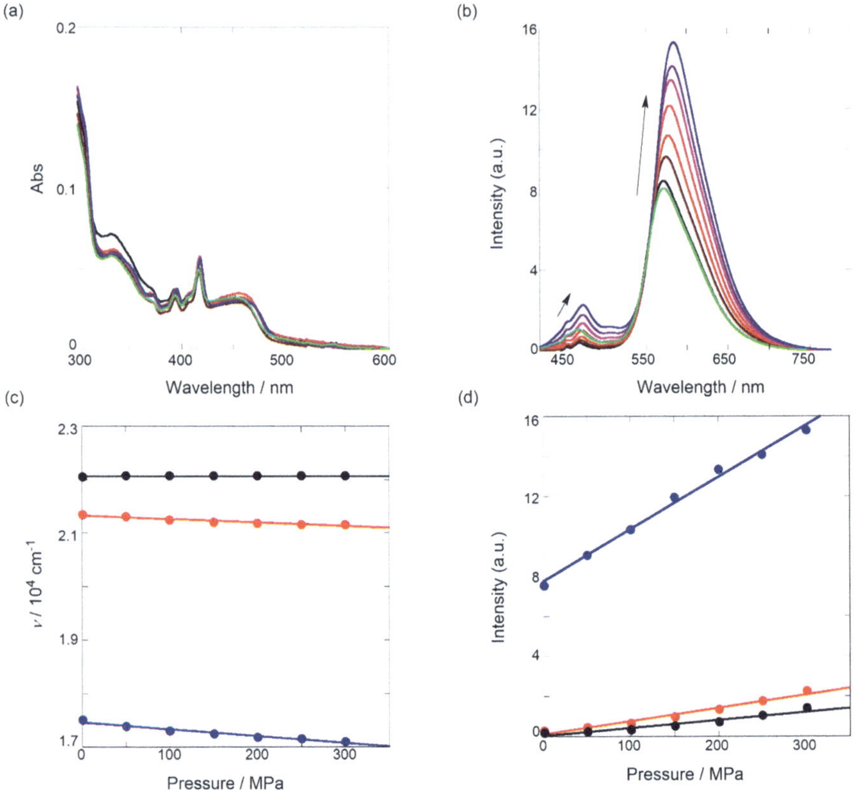

Fig. 4.18 (**a**) UV-vis absorption and (**b**) fluorescence spectra (excited at 400 nm) of **PTZ-DBPHZ** (126 μM) in MCH at room temperature under pressures of 0.1, 50, 100, 150, 200, 250, and 300 MPa (black to blue lines), along with the spectrum recorded after depressurization from 300 MPa (light green). All measurements were conducted using a high-pressure optical cell. (**c**) Pressure-dependent shifts of fluorescence maxima for the LE band (black; r = 0.612, slope = 0.0313 cm^{-1} MPa^{-1}), CT1 band (red; r = 0.959, slope = −0.691 cm^{-1} MPa^{-1}), and CT2 band (blue; r = 0.980, slope = −1.29 cm^{-1} MPa^{-1}). (**d**) Pressure dependence of fluorescence intensities at 452 nm (black; r = 0.967, slope = 0.00408 MPa^{-1}), 473 nm (red; r = 0.986, slope = 0.00660 MPa^{-1}), and 584 nm (blue; r = 0.996, slope = 0.0260 MPa^{-1}). (Reproduced with permission. Copyright 2019 John Wiley & Sons)

assigned to LE and a very short-lived species (τ_1 ~ 0.1–0.3 ns) attributable to CT1, with the latter becoming more prominent at longer monitoring wavelengths. The emission at 650 nm, arising from CT2, exhibited prompt fluorescence (τ ~ 10–11 ns) and a longer component (τ ~ 79–108 ns) ascribed to thermally activated delayed fluorescence (TADF). Notably, the relative TADF contribution (A_2/A_1) decreased steadily with increasing pressure, and eventually vanished above 300 MPa. This trend indicates that the viscosity increase under pressurization restricts donor rotation around the C-N bonds, thereby suppressing reverse intersystem crossing (rISC) and effectively eliminating the TADF channel, which underlies the observed ratiometric luminescence.

In contrast, the decay dynamics in MCH differed markedly. At ambient pressure, the LE and CT1 components showed similar lifetimes to those in toluene, but the A_2/A_1 ratio at 460 nm increased with pressure. Meanwhile, the CT2-associated decay at 570 nm displayed two components (τ_1 ~ 14–30 ns; τ_2 ~ 90–131 ns), and their ratio remained nearly constant throughout the pressure range, with both lifetimes gradually elongated. This behavior can be attributed to pressure-induced slowing of ISC/rISC processes, likely arising from an enlarged ΔE_{ST} in the nonpolar solvent. Collectively, these results highlight a striking contrast: in toluene, pressure induces ratiometric modulation through suppression of TADF, whereas in MCH the response is dominated by a monotonic increase of lifetimes without significant change in relative contributions.

Steady-state spectroscopy, time-correlated fluorescence decay, and quantum chemical calculations collectively provide insight into the pressure-dependent emission of **PTZ-DBPHZ** and the conformational dynamics in the ground and excited states (Fig. 4.19). DFT calculations identify four conformers, including *ax-eq*, *ax-ax* (*syn* and *anti*), and *eq-eq* with Boltzmann populations of 0.41:0.03:0.05:0.51 in toluene and 0.25:0.02:0.03:0.70 in MCH. Since *ax-ax* conformers are negligible, the discussion focuses on *ax-eq* and *eq-eq*. The minimal pressure dependence in the UV-vis spectra indicates that the ground-state conformer distributions remain nearly

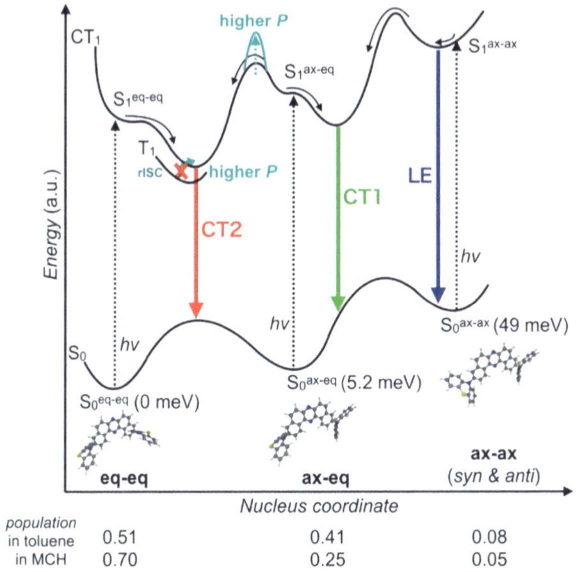

Fig. 4.19 Schematic representation of the ground and excited states of **PTZ-DBPHZ** in toluene. Values in parentheses denote the relative energies of the *ax-eq* and *ax-ax* conformers in the ground state with respect to the *eq-eq* conformer. (Reproduced with permission. Copyright 2019 John Wiley & Sons)

unchanged. In the excited state, however, interconversion occurs readily. In toluene, emission is dominated by the *eq-eq* conformer at 650 nm despite its smaller ground-state population, reflecting a low barrier to rearrangement, in line with related D-A luminophores. In MCH, CT emission from *eq-eq* conforms likewise dominates and persists under pressure, owing to their enhanced ground-state stability in nonpolar solvents. Differences in dipole moments (0.66 D for *eq-eq* and 5.77 D for *ax-eq*) further rationalize the solvent effect: nonpolar MCH (0.00 D) stabilizes *eq-eq*, whereas toluene (0.36 D) supports the coexistence of both conformers.

The application of hydrostatic pressure to toluene solutions increases viscosity and suppresses donor rotation around the C-N bond, thereby raising the barrier to interconversion and enhancing the relative contribution of *ax-eq* emission, as suggested by $\Delta V^{\neq}_{obs}$. In MCH, pressure primarily reduces nonradiative decay pathways, resulting in an overall increase in emission intensity without major changes in conformer distribution, while CT2 emission remains the dominant channel.

In conclusion, hydrostatic pressure exerts a profound yet nuanced influence on the excited-state properties of **PTZ-DBPHZ**, a conformationally flexible D-A-D compound. In polar toluene, pressure modulates the equilibria among LE, CT1, and CT2 emissions, yielding distinctive ratiometric responses governed by a crossover between thermodynamic and kinetic control regimes. In less polar MCH, pressure primarily suppresses nonradiative decay, enhancing all emissions without altering their balance. These insights establish pressure as a versatile stimulus for tuning the photophysics of mechanochromic luminophores and open avenues for the rational design of responsive materials based on conformational control in the excited state. By integrating spectroscopy, theory, and pressure-dependent methodology, this work advances fundamental understanding of mechanochromism and points toward innovative applications in sensing, imaging, and smart optical devices.

4.6 Ratiometric Pressure Sensing via S_1/S_2 Fluorescence Balance in Chiral Porphyrin Tweezers

A ratiometric chemosensing approach that compares two independent emissive channels offers a promising route to quantify pressure in solution while cancelling optical artifacts such as the shadowgraph effect (Fig. 4.20a, top). Ratiometric sensing based on the ratio of S_1 to S_2 fluorescence was investigated using chiral porphyrin tweezers, denoted **Por-Cy** and **Por-DPhEt**, in which two Zn-porphyrin units are connected by distinct flexible chiral linkers (Fig. 4.20b) (Ono et al., 2024). These systems were designed to couple the pressure-induced conformational equilibrium, specifically the open and closed (stacked) states of the tweezer scaffold (Fig. 4.20a, bottom). These two conformers were then linked to changes in the relative intensities of the S_1 and the higher energy S_2 emissions.

At 0.1 MPa, both tweezers displayed typical porphyrin Soret absorption at 427 nm and Q bands at 555 and 597 nm, together with S_1 fluorescence in the 550 to 750 nm range (Fig. 4.21a, b). Furthermore, an anti-Kasha fluorescence band was

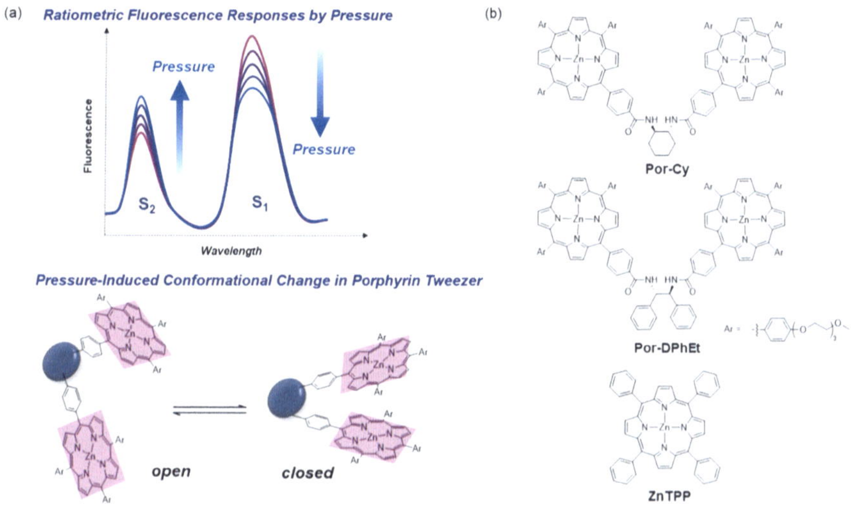

Fig. 4.20 (**a**) Conceptual diagram illustrating ratiometric fluorescence signaling based on the S_1 and S_2 excited states (top) and the pressure-induced open-closed conformational changes of the porphyrin tweezer (bottom). (**b**) Chemical structures of the porphyrin tweezers (**Por-Cy** and **Por-DPhEt**) and the reference compound **ZnTPP**. (Reproduced with permission. Copyright 2024 American Chemical Society)

observed near 434 nm for both tweezers under excitation in the Soret region (Fig. 4.22a, b, left panel) (Hirakawa et al., 2019). Excitation spectra monitored at this short wavelength matched the Soret absorption, confirming assignment of the short wavelength emission to S_2 fluorescence (Fig. 4.22a, b, right panel). Time-resolved measurements revealed that the S_2 emission decays on subnanosecond timescales, consistent with reported picosecond-scale S_2 lifetimes for porphyrins (Hirakawa et al., 2009). In contrast, S_1 decay in the open conformer was monoexponential with a lifetime of approximately 1.5 ns at atmospheric pressure. Chiral exciton coupling was clearly evident in CD spectra, with **Por-Cy** showing a strong bisignate Cotton effect and exciton coupling consistent with counterclockwise arrangement (Fig. 4.21c), and **Por-DPhEt** showing an exciton coupling pattern consistent with clockwise chromophore alignment (Fig. 4.21d).

DFT calculations of the linker rotation barriers indicated markedly different flexibilities for the two tweezers. The DPhEt alkane linker exhibited a lower rotational barrier, approximately 31 kJ mol^{-1} (Fig. 4.23b), while the cyclohexane-based Cy linker displayed a higher barrier, approximately 55 kJ mol^{-1}, owing to a half chair transition state (Fig. 4.23a). This difference suggested that **Por-DPhEt** should be more responsive to hydrostatic pressure-induced conformational changes than **Por-Cy**.

Concentration-dependent UV-vis measurements at 0.1 and 300 MPa confirmed that the pressure-dependent spectral changes observed at the concentrations employed do not arise from intermolecular aggregation but instead reflect intramolecular conformational equilibria. Pressurization produced reversible bathochromic

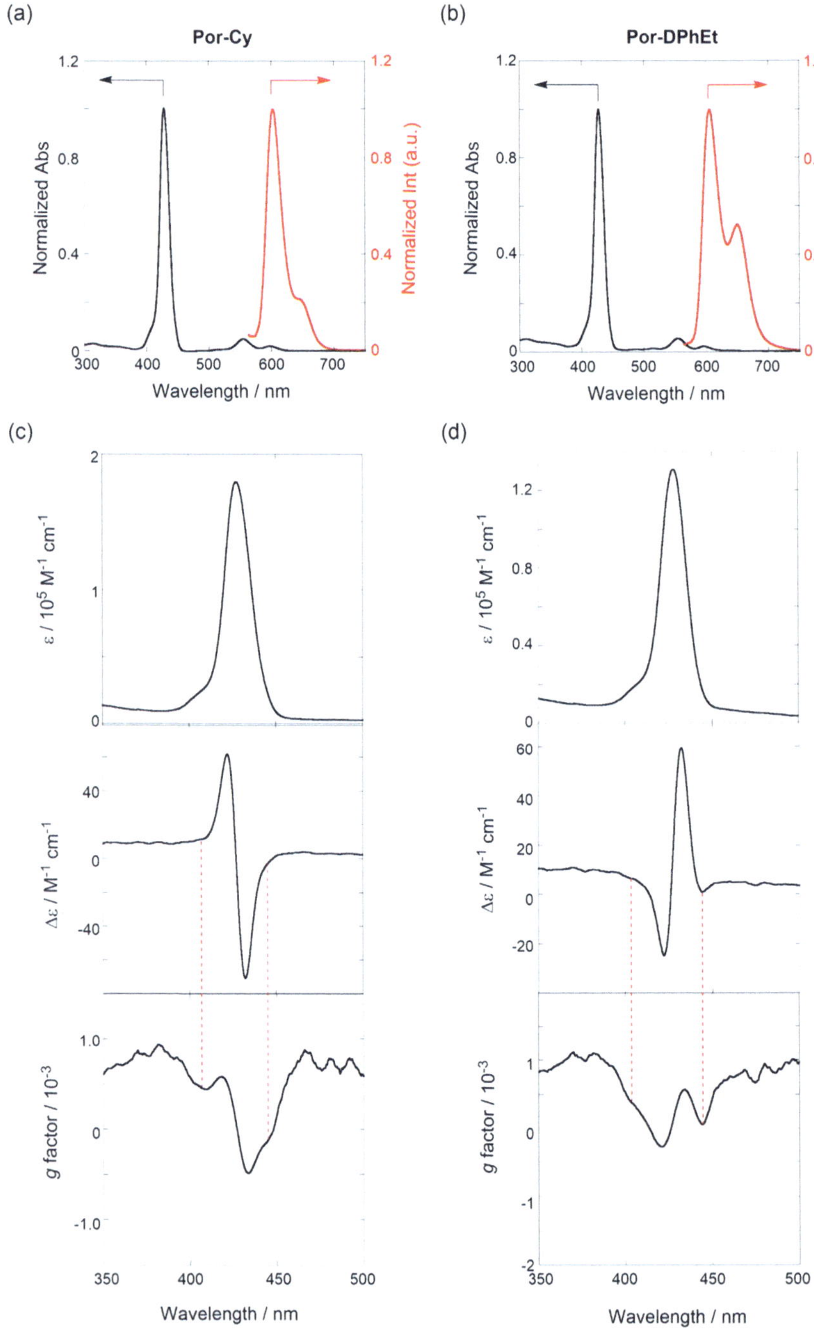

Fig. 4.21 (**a**, **b**) Normalized UV-vis absorption and fluorescence spectra (λ_{ex} = 554 nm) across the full wavelength range, and (**c**, **d**) UV-vis (top), CD (middle), and g-factor (bottom) spectra in the Soret region for (left) **Por-Cy** and (right) **Por-DPhEt** in toluene at 25 °C using a 1-cm quartz cell. (Reproduced with permission. Copyright 2024 American Chemical Society)

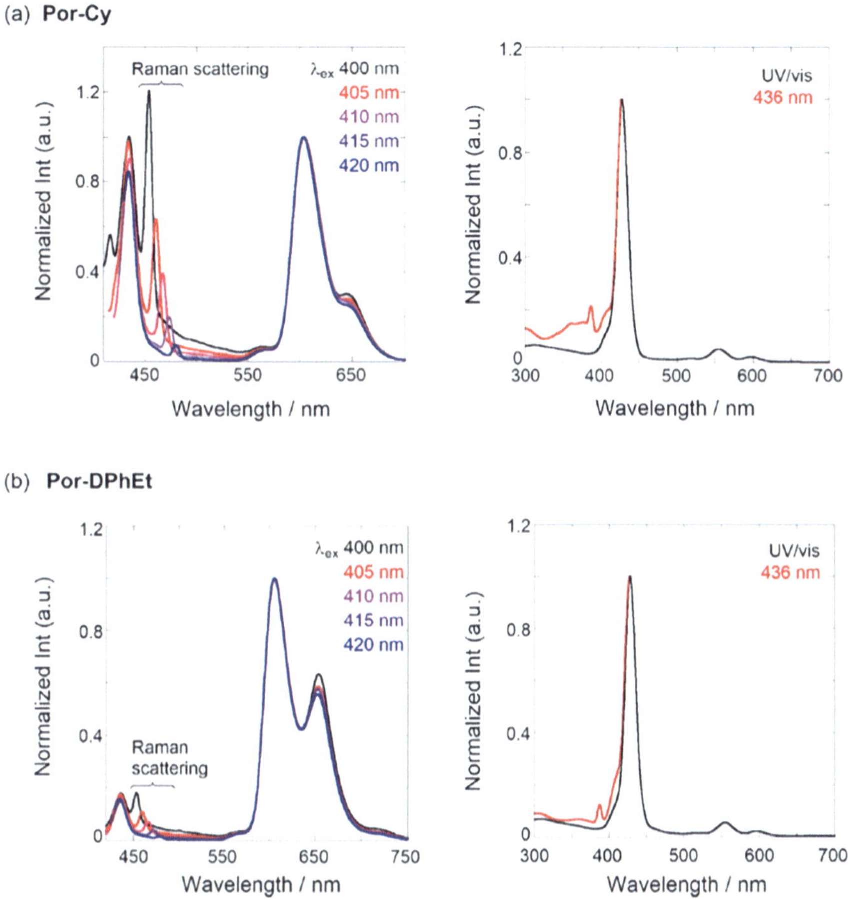

Fig. 4.22 Normalized fluorescence (λ_{ex} = 400–420 nm, left) and excitation (λ_{obs} = 436 nm, right) spectra of (**a**) **Por-Cy** and (**b**) **Por-DPhEt** in toluene at 25 °C using a 1-cm quartz cell. (Reproduced with permission. Copyright 2024 American Chemical Society)

shifts of the Soret band, with −0.917 cm^{-1} MPa^{-1} for **Por-Cy** (Fig. 4.24a, left) and −0.861 cm^{-1} MPa^{-1} for **Por-DPhEt** (Fig. 4.24b, left). These shifts are attributed to increased solvent polarizability and enhanced solvation stabilization of the π* orbitals upon pressurization. In addition to steady shifts, pressure-induced hyperchromic effects were observed, reflecting increased effective local concentration with solvent pressurization.

Fluorescence responses to increasing hydrostatic pressure differed between the two tweezers. **Por-Cy** exhibited a gradual decrease in intensity for both S_1 and S_2 emissions upon pressurization (Fig. 4.24a, right). Conversely, **Por-DPhEt** exhibited a decrease in S_1 intensity and an increase in S_2 intensity with increasing pressure (Fig. 4.24b, right). Both systems displayed isoemissive points in their fluorescence spectra, consistent with an equilibrium between two emissive conformers,

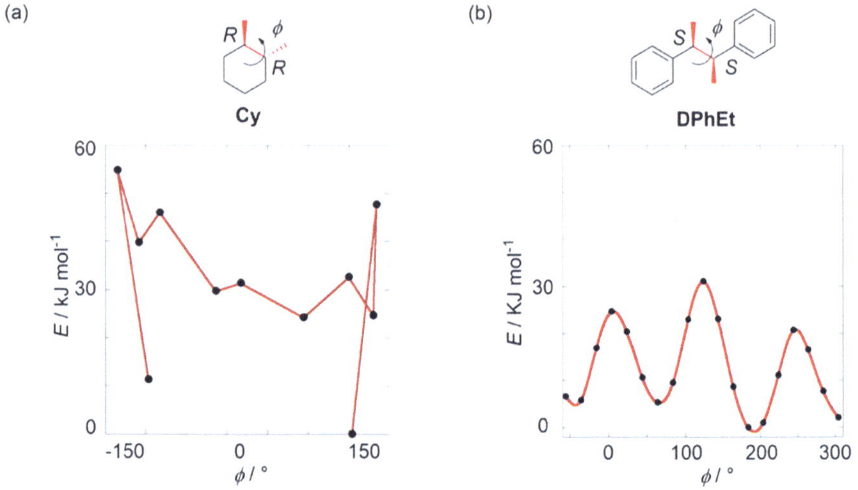

Fig. 4.23 Potential energy profiles for dihedral angle variations of (**a**) Cy and (**b**) DPhEt calculated at the ωB97xd/6-31G(d,p) level; the lowest-energy conformer in each scaffold was set to 0 kJ mol^{-1}. (Reproduced with permission. Copyright 2024 American Chemical Society)

interpreted as open and closed forms. Time-resolved measurements under pressure supported this assignment. At atmospheric pressure, S_1 fluorescence decays were dominated by a single long-lived component attributable to the open conformer. Upon pressurization, decay profiles showed two exponential components: a longer-lived species at approximately 1.5 ns and a short-lived component at about 0.2–0.3 ns. The latter was assigned to closed, stacked conformers in which stacking promotes radiationless deactivation. The increasing fraction of the short-lived component with pressure mirrored the S_1 quenching behavior.

CD spectra provided further evidence for pressure-modulated open-closed equilibria. Subtle shoulders and secondary bands in the Soret region were interpreted in terms of *H*-type and *J*-type exciton-coupled closed species, in addition to the open exciton-coupled signals (Fig. 4.21c, d) (Hestand & Spano, 2018). Quantitative analysis of the FWHM of the Soret band revealed opposite trends for the two systems upon pressurization. **Por-Cy** showed band broadening consistent with the development of a short-wavelength, *H*-type component (Fig. 4.25a), whereas **Por-DPhEt** exhibited band narrowing consistent with the growth of a longer-wavelength, *J*-type component (Fig. 4.25b). Thus, hydrostatic pressure shifts the conformational equilibrium in both systems, while the character of the stacked state, *H* or *J*, is dictated by linker structure and flexibility.

To elucidate the behavior of S_2 emission under pressure, **ZnTPP** was examined as a monomeric reference. In **ZnTPP**, S_1 fluorescence intensity increased with pressure while S_2 decreased (Fig. 4.26a). The pressure-induced enhancement of S_1 was attributed to inhibited collisional deactivation by the more viscous compressed solvent (Fig. 4.26b). The decrease of S_2 with pressure suggested enhanced radiationless deactivation or inhibition of fluorescence at the higher excited state, likely

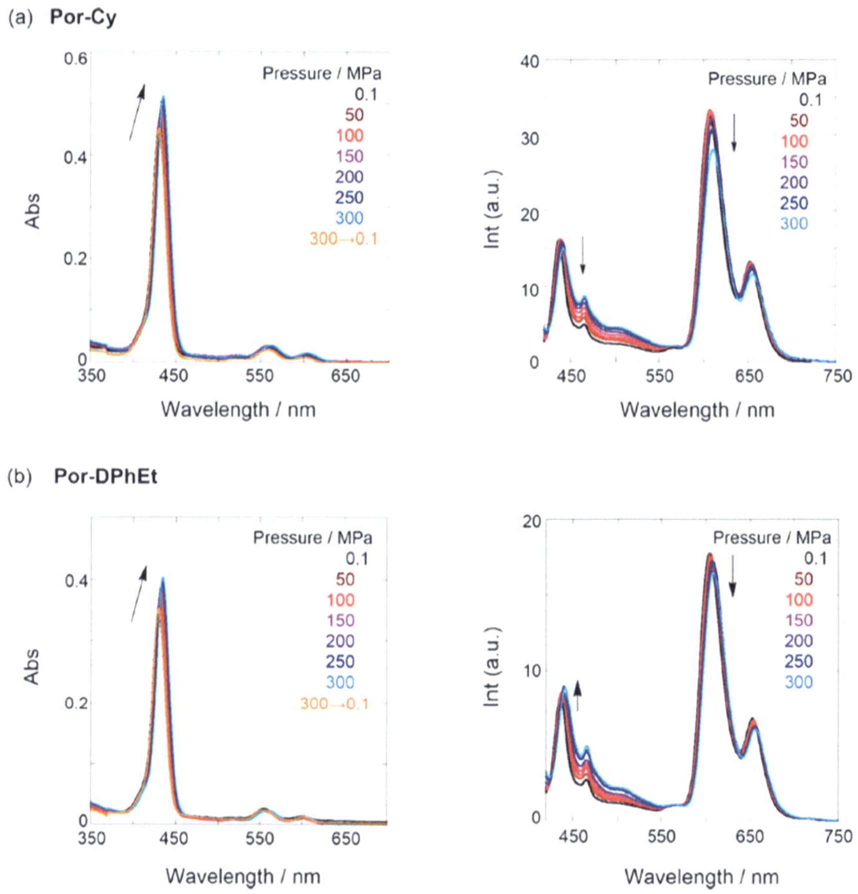

Fig. 4.24 UV-vis (left) and fluorescence (λ_{ex} = 410 nm, right) spectra of (**a**) **Por-Cy** (12.1 μM) and (**b**) **Por-DPhEt** (9.7 μM) in toluene at 0.1, 50, 100, 150, 200, 250, and 300 MPa (black to sky-blue lines) recorded at room temperature in a high-pressure cell. The orange line corresponds to the UV-vis spectrum at 0.1 MPa after depressurization from 300 MPa. (Reproduced with permission. Copyright 2024 American Chemical Society)

linked to pressure-induced changes in solvent properties and excited-state π electron distributions. The S_2 responses of the tweezers were rationalized on this basis: the *H*-stacked species in **Por-Cy** followed the **ZnTPP** trend of S_2 quenching with pressure, whereas the *J*-stacked species in **Por-DPhEt** presented an altered excited-state microenvironment such that S_2 emission increased under pressurization.

Quantification of pressure sensitivity was achieved using a ratiometric response factor γ, defined as the ratio of S_1 to S_2 fluorescence intensity as a function of pressure. Both tweezers exhibited linear γ versus pressure relationships with high correlation coefficients (Fig. 4.27a, b). **Por-DPhEt** displayed a slope γ of 1.6×10^{-3} MPa^{-1}, approximately 2.7 times larger than that of **Por-Cy**

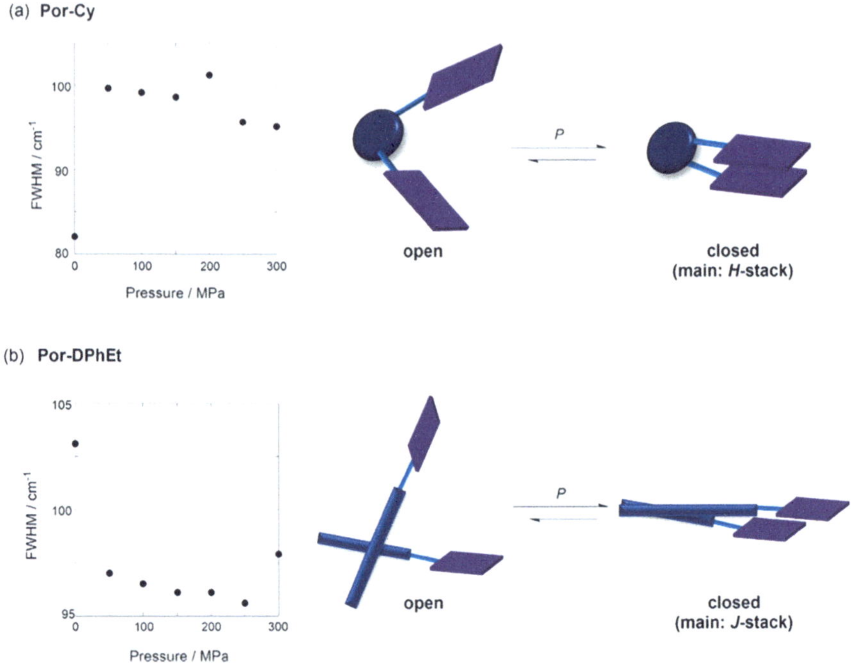

Fig. 4.25 Pressure dependence of the FWHM of the Soret band (left) and schematic representations of the corresponding open-closed conformational change (right) of (**a**) **Por-Cy** and (**b**) **Por-DPhEt**. (Reproduced with permission. Copyright 2024 American Chemical Society)

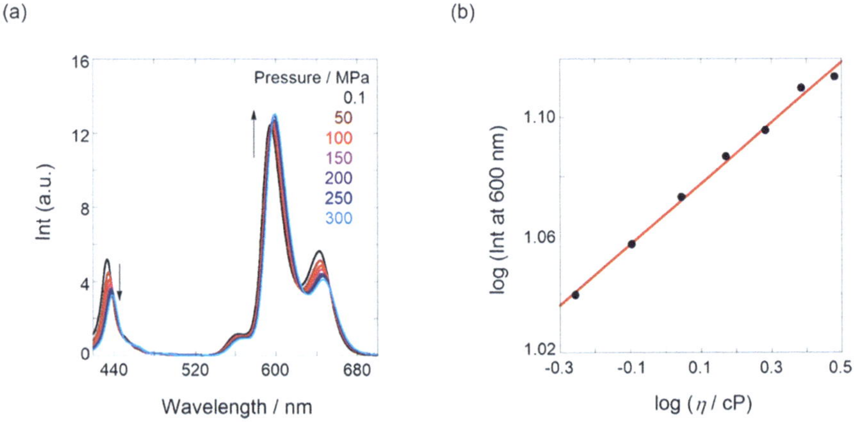

Fig. 4.26 (**a**) Fluorescence spectra (λ_{ex} = 410 nm) of **ZnTPP** (2.9 μM) in toluene at 0.1, 50, 100, 150, 200, 250, and 300 MPa (from black to sky-blue lines) recorded at room temperature in a high-pressure cell. (**b**) Logarithmic plot of the fluorescence intensity at 600 nm obtained from (**a**) as a function of viscosity (η), analyzed using the Förster-Hoffmann relation. (Reproduced with permission. Copyright 2024 American Chemical Society)

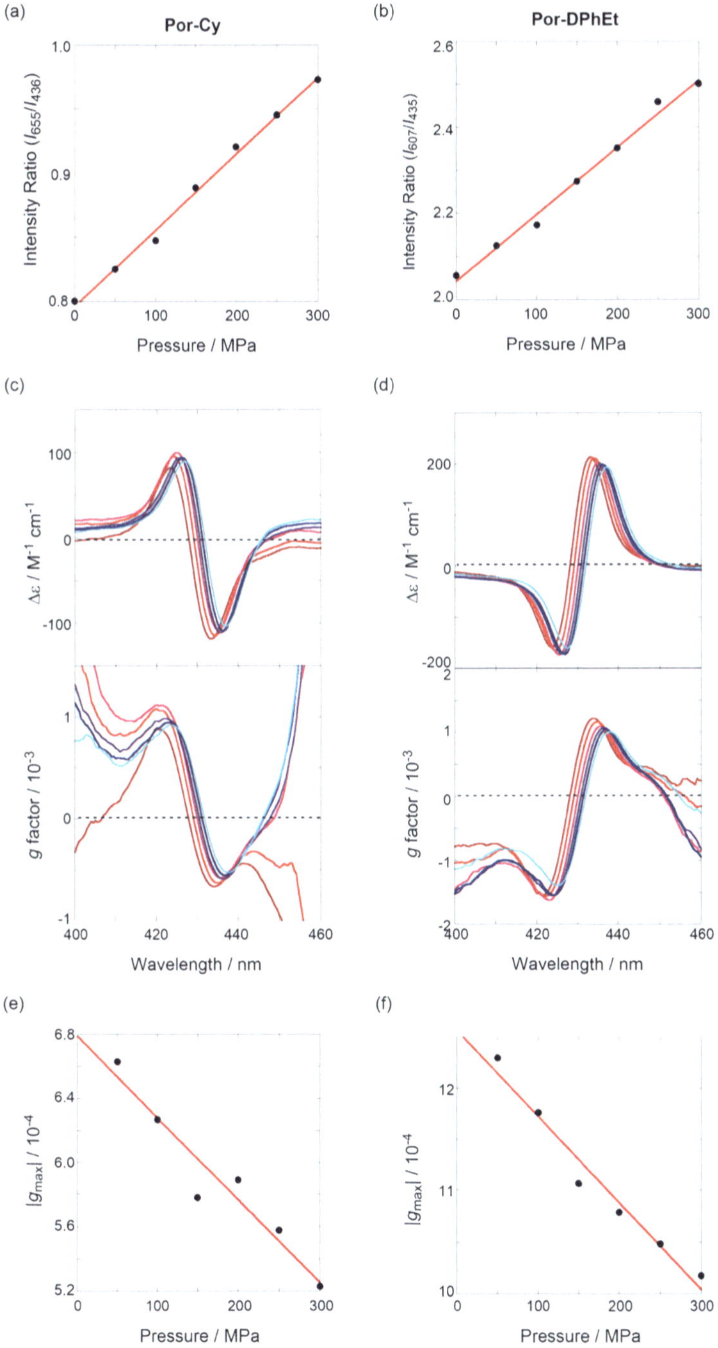
(a)
Por-Cy
Intensity Ratio (I_{655}/I_{436})
Pressure / MPa
(b)
Por-DPhEt
Intensity Ratio (I_{607}/I_{435})
Pressure / MPa
(c)
Δε / M⁻¹ cm⁻¹
g factor / 10⁻³
Wavelength / nm
(d)
Δε / M⁻¹ cm⁻¹
g factor / 10⁻³
Wavelength / nm
(e)
$|g_{max}|$ / 10⁻⁴
Pressure / MPa
(f)
$|g_{max}|$ / 10⁻⁴
Pressure / MPa

($\gamma = 5.9 \times 10^{-4}$ MPa^{-1}). The enhanced sensitivity of **Por-DPhEt** is consistent with its lower rotational barrier and greater scaffold flexibility, permitting a larger pressure-induced shift in conformational equilibrium. Hydrostatic pressure-dependent CD spectroscopy provides valuable mechanistic insight into the structural flexibility of the linker. As shown in Fig. 4.27c, d, the first Cotton effect at longer wavelengths gradually diminished with increasing pressure for both tweezers. The *g*-based response factor (γ_{gabs}) was determined from the plots of $|g_{max}|$ versus pressure (Fig. 4.27e, f), with **Por-DPhEt** exhibiting a γ_{gabs} value 1.6 times larger (-8.4×10^{-7} MPa^{-1}) than that for **Por-Cy** (-5.1×10^{-7} MPa^{-1}). Such a finding indicates that the enhanced pressure responsiveness originates from the ground-state chiral conformational change, reflecting the greater structural flexibility of the **Por-DPhEt** scaffold.

In conclusion, ratiometric sensing based on the S_1/S_2 fluorescence ratio in porphyrin tweezers provides a practical and novel mechanism to quantify hydrostatic pressure in solution. The sensing operates through pressure-induced open-to-closed conformational switching within a single molecule, which shifts the balance between radiative and nonradiative decay at the S_1 and S_2 states. Linker flexibility strongly controls sensitivity and stacking geometry. The more flexible DPhEt linker gives a larger pressure response factor than the less flexible Cy linker, and promotes *J*-type stacking that enhances the S_2 emission. On the other hand, the less flexible **Por-Cy** exhibits *H*-type stacking and pressure-induced quenching of S_2 emission. Because the S_1/S_2 ratiometry compensates for pressure-induced optical artifacts such as refractive index gradients, and because scaffold design enables tuning of both sensitivity and response direction, S_2-based ratiometric chemosensing offers a clerobustar platform for the development of smart pressure-responsive sensors.

4.7 Solvation-Driven Conformational Switching of Propeller-Chiral Hexaarylbenzenes Under Hydrostatic Pressure

Hydrostatic pressure was employed as a mechanistic probe to quantify how propeller-chiral hexaarylbenzenes respond to changes in local solvation and to map the coupled relationships among intramolecular geometry, cavity solvation, and chiroptical output. Two substituted hexaarylbenzenes, denoted **H2** and **H6** (Fig. 4.28)

Fig. 4.27 (Left) **Por-Cy** and (right) **Por-DPhEt**: (**a**, **b**) plots of fluorescence intensity ratios as a function of pressure (*r* = 0.997 (left) and 0.995 (right)); (**c**, **d**) pressure-dependent CD (top) and *g*-factor (bottom) spectra in the Soret region in toluene at 50, 100, 150, 200, 250, and 300 MPa (from brown to sky-blue lines); and (**e**, **f**) plots of $|g_{max}|$ values for the first Cotton effect obtained from (**c**, **d**) versus pressure (*r* = 0.963 (left) and 0.983 (right)). (Reproduced with permission. Copyright 2024 American Chemical Society)

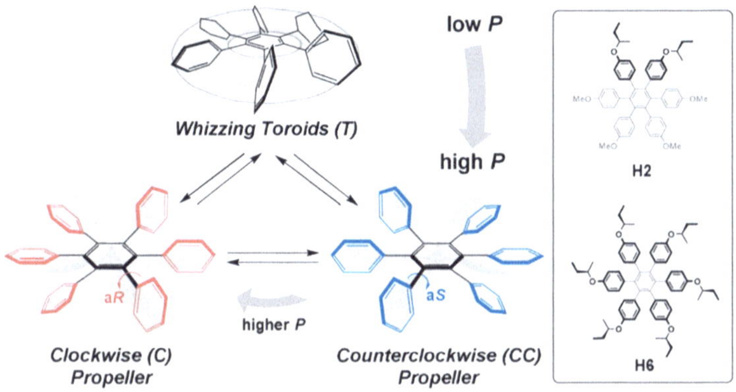

Fig. 4.28 Conformational dynamics of hexaarylbenzenes (HABs) involving toroidal (*T*) motions and propeller-type chiral forms (*C* and *CC*). Insets: molecular structures of **H6** and **H2** examined in this study. (Reproduced with permission. Copyright 2019 John Wiley & Sons)

(Kosaka et al., 2016, 2018), were examined in MCH and hexane through pressure-resolved UV-vis absorption and CD spectra recorded up to 320 MPa (Kosaka et al., 2019). Spectroscopic observables were interpreted within a minimal three-state framework that recognizes clockwise (*C*) and counterclockwise (*CC*) propeller minima and an ensemble of intermediate toroidal (*T*) geometries.

Substantial bathochromic shifts of the split ^1L_b absorption band and concurrent increases in absorption intensity occurred with applied pressure (Fig. 4.29). Pressure coefficients for those shifts exceeded values for simple chiral biaryls, indicating an amplified cooperative response of the six peripheral aromatic blades. For example, **H2** in MCH showed low-pressure slopes near −2.1 and −2.9 cm^{-1} MPa^{-1} for the two split peaks, whereas typical biaryls yield slopes of −0.4 to −0.6 cm^{-1} MPa^{-1}. **H6** displayed comparable pressure sensitivity, while **H6** in linear hexane exhibited even steeper low-pressure coefficients (Fig. 4.30). The steeper coefficients for hexane are consistent with easier solvent penetration into the intramolecular cavity under pressurization in linear solvents relative to bulkier cyclic solvents.

Relative intensity changes between split peaks at 290 and 270 nm evolved systematically with pressure in CD spectra (Fig. 4.29). This behavior is consistent with enhanced propeller-like excitonic coupling relative to toroidal character upon initial pressurization. Deviations from strict isodichroic behavior indicated simultaneous modulation of the solvation shell rather than a simple two-state interconversion. CD spectra corroborated UV-vis trends and revealed solvent-specific complexity. In MCH, anisotropy increased progressively in the low-to-middle pressure domain and then leveled off or slightly decreased at higher pressures, consistent with the initial desolvation-driven conversion of *T* to propeller followed by modest *C-CC* redistribution.

Nonmonotonic CD behavior was observed for **H6** in hexane, with anisotropy rising substantially at intermediate pressures and declining at higher pressures. That pattern reflects competing effects: geometric stabilization of the propeller blade tilt that enhances chiroptical response at moderate pressure, followed by a

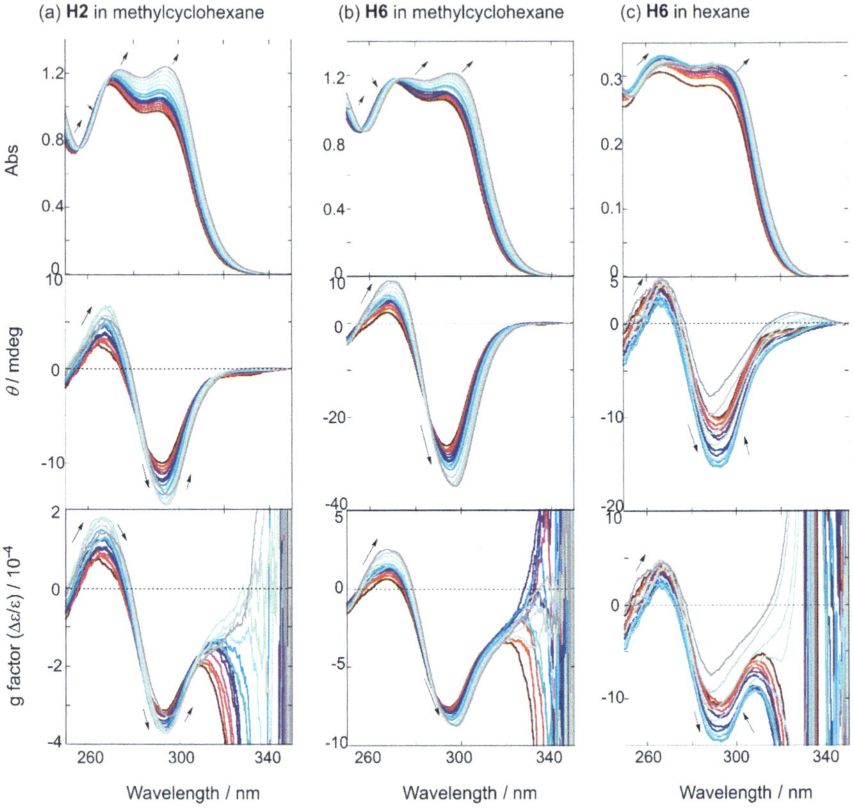

Fig. 4.29 Pressure-dependent UV-vis (top), CD (middle), and anisotropy (g) factor (bottom) spectra of (**a**) **H2** and (**b**) **H6** in MCH (130 μM), and (**c**) **H6** in hexane (33 μM) at ambient temperature. Measurements were performed under applied pressures of 10, 20, 30, 40, 50, 60, 70, 80, 90, 100, 120, 160, 200, 240, 280, and 320 MPa. (Reproduced with permission. Copyright 2019 John Wiley & Sons)

pressure-driven population shift toward the energetically less favored *CC* diastereomer when solvent-specific differential volumes make *CC* comparatively more stable under extreme pressurization. The observed spectral inflections near 100 MPa mark a crossover where solvent-mediated fixation of blade geometry becomes kinetically or thermodynamically dominant.

A geometric descriptor based on the dihedral-angle dependence of electronic transition probabilities provided an internally consistent picture of pressure-induced distortions. The difference between ground- and excited-state dihedral angles, $\theta_g - \theta_e$, decreased systematically with pressure for both **H2** and **H6** in MCH at approximately 5° per 100 MPa—substantially larger than the 0.9–1.4° per 100 MPa reported for representative biaryls. This pronounced angular sensitivity reflects pressure-induced blade flattening in the ground state and confirms that hexaarylbenzenes behave as mechanically compliant motifs, whose electronic couplings are

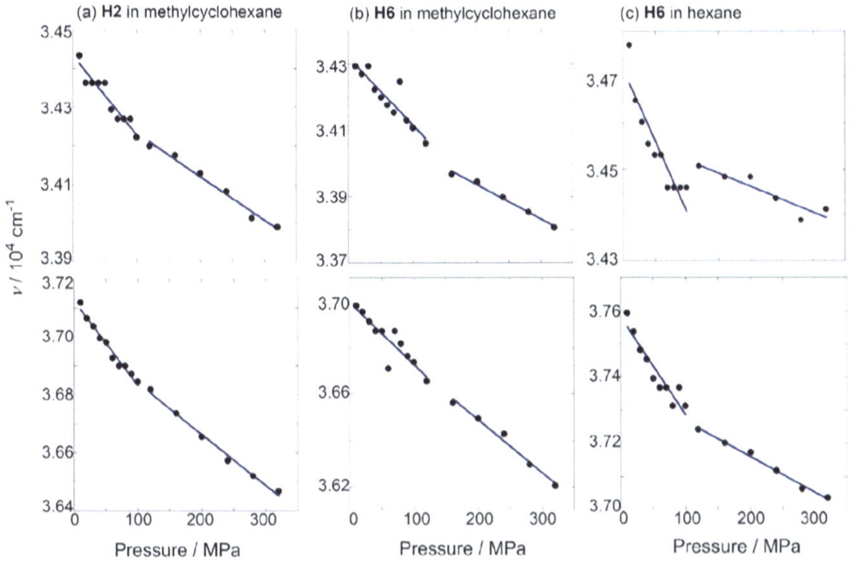

Fig. 4.30 Pressure dependence of the absorption maxima of the split ^1L_b band for (**a**) **H2** and (**b**) **H6** in MCH, and (**c**) **H6** in hexane. (Reproduced with permission. Copyright 2019 John Wiley & Sons)

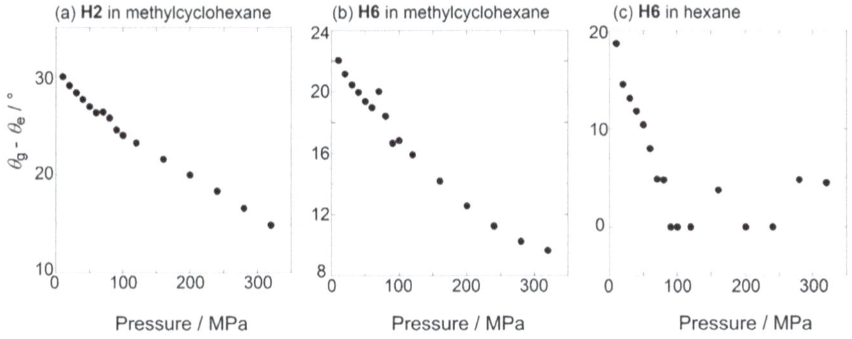

Fig. 4.31 Pressure dependence of the differential dihedral angle between the S_0 and S_1 states ($\theta_g - \theta_e$) for (**a**) **H2** and (**b**) **H6** in MCH, and (**c**) **H6** in hexane. (Reproduced with permission. Copyright 2019 John Wiley & Sons)

tightly linked to blade orientation. For **H6** in hexane, $\theta_g - \theta_e$ approached near zero near 100 MPa, indicating conformational fixation above that threshold and explaining abrupt spectral changes occurring at similar pressures (Fig. 4.31).

The differential molecular volumes play a key role in governing the pressure dependence of the (chir)optical properties of HABs by constraining the equilibrium between the toroidal and propeller conformers. Figure 4.32 presents the pressure dependence of the g factors at two wavelengths within the ^1L_b band for **H2** and **H6**.

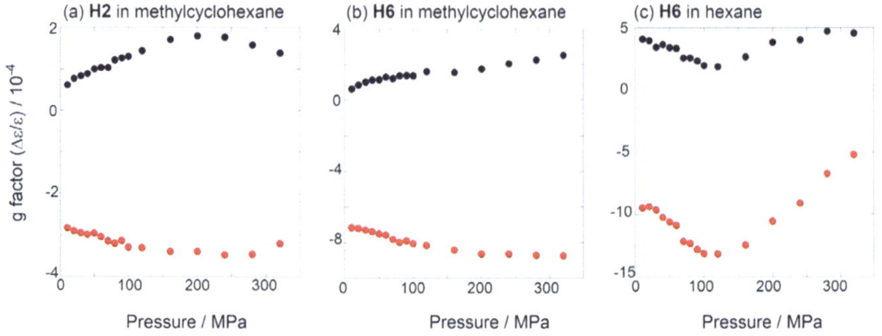

Fig. 4.32 Pressure dependence of the g factors for (**a**) **H2** and (**b**) **H6** in MCH, and (**c**) **H6** in hexane, monitored at 267 nm (black) and 298 nm (red). (Reproduced with permission. Copyright 2019 John Wiley & Sons)

In MCH, both compounds exhibited similar pressure responses: the g-factor intensity increased progressively at low to moderate pressures, while the changes became less pronounced (or slightly reversed for **H2**) at higher pressures. The effective differential volumes ($\Delta V_{CC\text{-}C}$ and $\Delta V_{T\text{-}C}$) for **H2** and **H6** were estimated under the following assumptions: (1) three possible conformers (toroidal (T) and propeller (clockwise and counterclockwise, C/CC)) exist, and (2) their g factors are 0, $+g_0$ (+0.003), and $-g_0$ (−0.003), respectively. Using the relation $-RT\ln K = \Delta F + P\Delta V$, that slopes of the pressure-dependent equilibria yield effective differential volumes. Remarkably large values of $\Delta V_{T\text{-}C}$ were obtained in solution; approximately +42.6 and +41.9 $cm^3\ mol^{-1}$ for **H2** and **H6**, respectively, in MCH, and +72.2 $cm^3\ mol^{-1}$ for **H6** in hexane. Conversely, static van der Waals volume differences computed for optimized gas-phase geometries yielded $\Delta V_{T\text{-}C}$ ~ +1.3 $cm^3\ mol^{-1}$ and $\Delta V_{CC\text{-}C}$ ~ +0.13 $cm^3\ mol^{-1}$. This large disparity highlights that the experimentally-observed differential volumes are dominated by solvation and desolvation of the intramolecular cavity rather than intrinsic molecular volume changes.

Molecular interpretation identifies the toroidal ensemble as effectively larger because an extended solvent shell occupies the intramolecular void, whereas propeller conformers exclude this solvent shell and therefore present a smaller effective volume. Solvent morphology and size are thus decisive: linear hexane penetrates and is removed more readily from the cavity under pressure than MCH, explaining the substantially larger $\Delta V_{T\text{-}C}$ observed for **H6** in hexane. Differential volumes between C and CC remained small in MCH (+0.17 and +0.32 $cm^3\ mol^{-1}$ for **H2** and **H6**, respectively), consistent with the near cancellation of solvation contributions when the two propellers present similar cavity exposure. Notably, the sign of $\Delta V_{CC\text{-}C}$ inverts for **H6** in hexane (−7.7 $cm^3\ mol^{-1}$), accounting for the observed reduction in CD intensity at high pressure in hexane by favoring the CC diastereomer under pressurization, a consequence of differing solvation modes around the two propeller forms.

In conclusion, this study elucidates how pressure modulates the conformational equilibria of hexaarylbenzenes (HABs) through the interplay of solvation thermodynamics and mechanical effects. The pronounced sensitivity of differential molecular volumes to solvent identity highlights that both enthalpic and entropic factors of solvation, governed by subtle variations in solvent shape and packing, are crucial in determining the pressure response. Pressure acts not only as a compressive force but also as a perturbation to solvation thermodynamics, coupling desolvation with mechanical fixation to drive conformational switching between toroidal and propeller forms. Initial cavity contraction promotes desolvation and enhances chiroptical activity through increased excitonic coupling, while further pressurization leads to solvent-specific stabilization of distinct diastereomers. These findings establish hexaarylbenzene frameworks as promising candidates for piezo-responsive optical materials. Their large, solvent-dependent $\Delta V_{T\text{-}C}$ values indicate that small changes in hydrostatic pressure or molecular packing can produce significant variations in electronic and chiroptical properties. Rational molecular design that optimizes cavity accessibility, substituent bulkiness, and solvent compatibility will allow precise control over conformational bias and pressure sensitivity, enabling applications in pressure sensing and actuation. Furthermore, the combined use of pressure-dependent spectroscopy and volumetric analysis provides a general approach for disentangling solvation effects from intrinsic molecular geometry in flexible, cavity-bearing systems, offering a conceptual foundation for the design of mechanically responsive molecular materials.

4.8 Solvent-Dependent Chiroptical Switching of Molecular Universal Joints Under Pressure

Planar-chiral cyclophano-pillar[5]arenes (cyclophano-P[5]s) bearing a fused oligo(oxyethylene) or polymethylene subring—hereafter referred to as molecular universal joints or **MUJs** (Yao et al., 2017)—exhibit an equilibrium between subring-included (in) and subring-excluded (out) conformers whose populations respond sensitively to hydrostatic pressure (Fig. 4.33) (Yao et al., 2021). Pressure-dependent UV-vis and CD spectroscopy across a broad solvent set reveals that modulation of the in–out equilibrium produces unusually large chiroptical changes, including an unprecedented pressure-driven inversion of planar chirality and, for the **MUJ** bearing a dodecamethylene subring, the largest reported single-sample change in the *g* factor. The magnitude and sign of chiroptical response under pressurization depend strongly of both subring identity and solvent identity, which together determine the balance among subring self-inclusion, solvent inclusion into the P[5] cavity, and solvation of the excluded subring.

Optical resolution of racemic **MUJ1**-**MUJ3** furnished enantiomerically enriched samples for spectroscopic interrogation under hydrostatic pressures up to 320 MPa using a high-pressure vessel equipped with birefringence-free diamond windows. Solvents spanning nonpolar hydrocarbons to polar aprotic media were examined, including hexane, MCH, carbon tetrachloride, THF, ethyl acetate, chloroform,

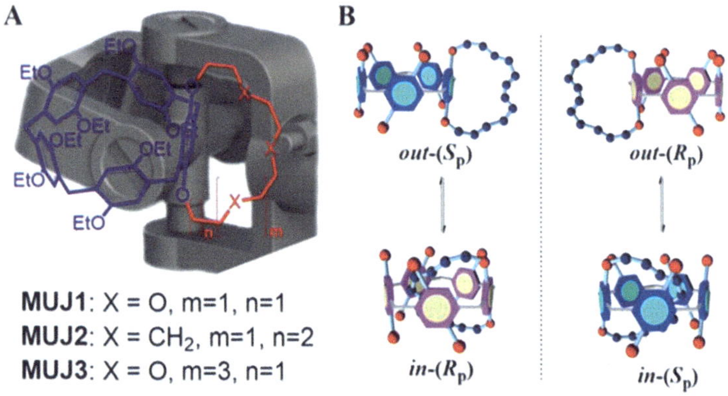

Fig. 4.33 Schematic of the chemical structures and chirality switching of molecular universal joints (**MUJs**). (Left) Pillar[5]arene-based **MUJs**. (Right) Illustrations of the in-out equilibrium between the enantiomeric pairs of **MUJ2**. (Reproduced with permission. Copyright 2021 Royal Society of Chemistry)

DCM, and acetonitrile. CD spectra and pressure-normalized g factors = $\Delta\varepsilon/\varepsilon$ were monitored as functions of pressure. Reversible recovery of initial spectra upon pressure release, without hysteresis, established that the observed changes reflect reversible conformational and solvation equilibria rather than chemical decomposition or irreversible conformational locking.

The behavior of **MUJ2** highlights the solvent-dependent richness of pressure responses. In acetonitrile, the g factors at 258 and 310 nm underwent sign inversion with pressure, shifting from −0.0035 to +0.0039 at 258 nm and from −0.0016 to +0.0034 at 310 nm upon raising pressure from 0.1 to 320 MPa (Fig. 4.34a). Net changes in g reached 0.0074 at 254 nm and 0.0050 at 310 nm, exceeding previously reported pressure-induced g changes for related supramolecular systems. In contrast, ethyl acetate produced only a modest initial increase in g_{310} followed by a slow decrease (Fig. 4.34b), yielding a far smaller net change of 0.0007 under the same pressure window. **MUJ1** and **MUJ3** exhibited distinct solvent-dependent patterns, with **MUJ3** often resembling **MUJ2** in sign but differing in magnitude, and **MUJ1** showing distinct solvent-specific trends attributable to its smaller polyether subring (Fig. 4.35).

The semi-quantitative descriptor, pressure susceptivity $\Delta g/\Delta P = (g_{310}{}^{P\max} - g_{310}{}^{P\min})/(P_{\max} - P_{\min})$, captures the overall g responsiveness across the applied pressure range despite nonuniform pressure dependence. Large positive susceptivities indicate a pressure-driven shift toward the in-(R_p) conformer, while negative values indicate shift toward out-(S_p). **MUJ2** displayed particularly large positive $\Delta g/\Delta P$ values in DCM and acetonitrile, +5.9 and +15.6 TPa^{-1}, respectively, consistent with the observed pressure-driven sign inversions in these solvents. In many less polar solvents, the $\Delta g/\Delta P$ values were smaller or even negative, implying either weaker driving toward the in-conformer or competing mechanisms that favor the out-conformer upon pressurization.

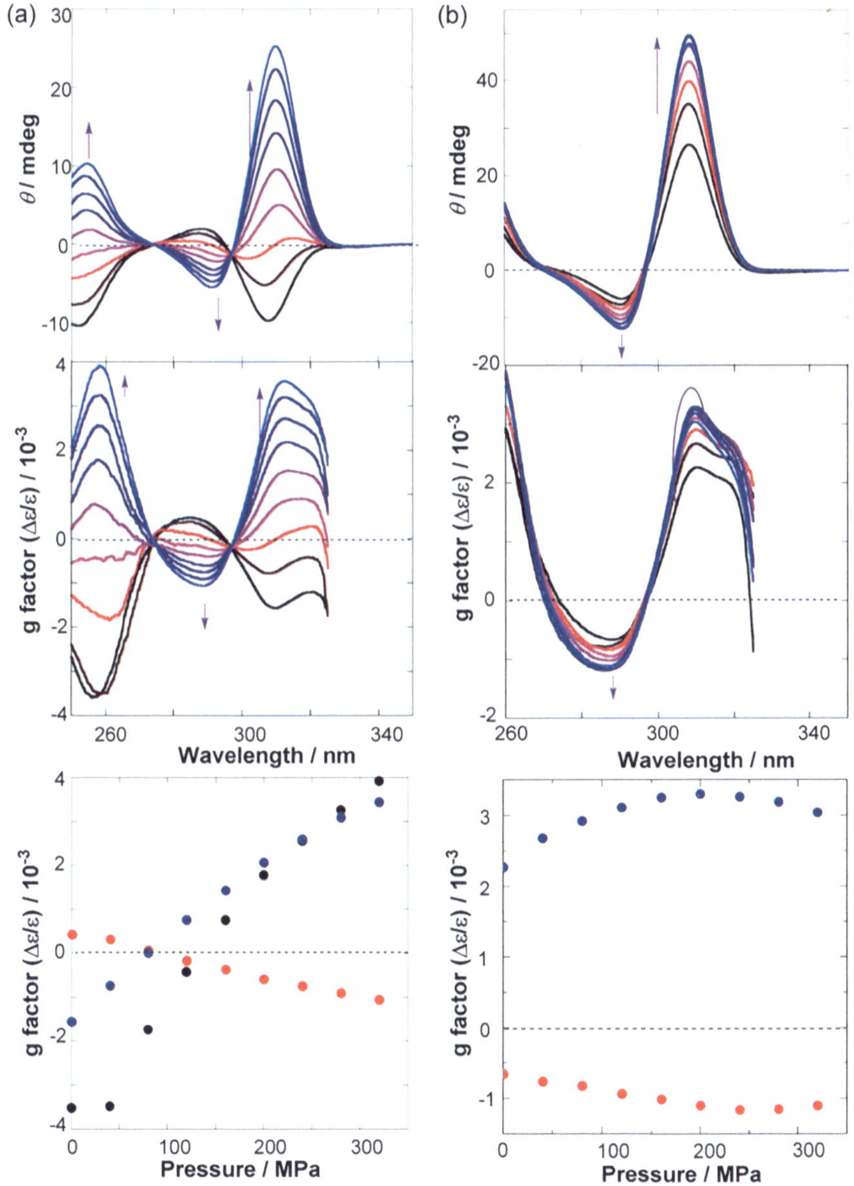

Fig. 4.34 Pressure-dependent spectral analysis of in-(R_p)/out-(S_p)-**MUJ2**. (**a**) CD (top), g factor (middle), and pressure-dependent g-factor plots (bottom) monitored at 310 nm (blue), 288 nm (red), and 258 nm (black) for in-(R_p)/out-(S_p)-**MUJ2** (92 μM) in acetonitrile at room temperature. (**b**) Corresponding CD (top), g-factor (middle), and pressure-dependent g-factor plots (bottom) at 309 nm (blue) and 288 nm (red) for in-(R_p)/out-(S_p)-**MUJ2** (136 μM) in ethyl acetate at room temperature. Applied pressures: 0.1, 40, 80, 120, 160, 200, 240, 280, and 320 MPa (from black to light blue). (Reproduced with permission. Copyright 2021 Royal Society of Chemistry)

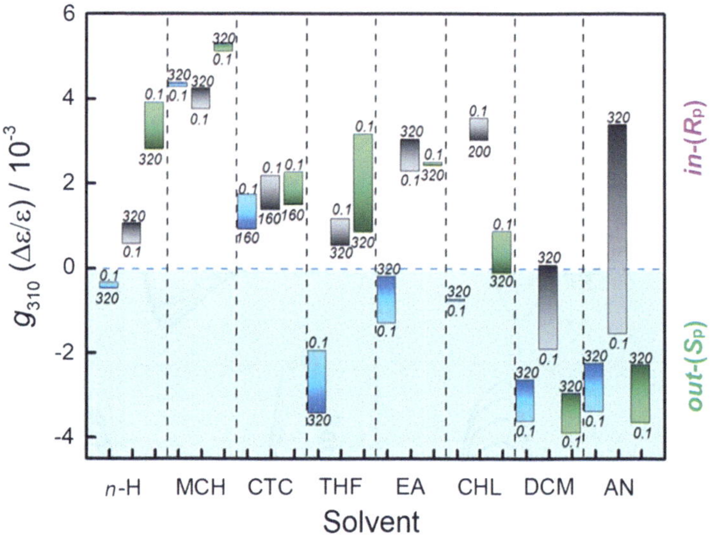

Fig. 4.35 Pressure dependence of the *g* factor for **MUJs**. Pressure-induced variations of the *g* factor at 310 (±1) nm (g_{310}) for in-(R_p)/out-(S_p)-**MUJ1** (blue), **MUJ2** (gray), and **MUJ3** (green) in various solvents such as hexane (*n*-H), MCH, carbon tetrachloride (CTC), THF, ethyl acetate (EA), chloroform (CHL), DCM, and acetonitrile (AN) at room temperature. (Reproduced with permission. Copyright 2021 Royal Society of Chemistry)

Evaluations of the intrinsic molecular volumes by DFT at the B3LYP/6-31G(d) level produced van der Waals volumes for the in- and out-conformers that differ by only 0.26–0.36 Å^3, a negligible fraction of the total MUJ volume. Such small static volume differences preclude purely steric volume changes as the primary driver for pressure-induced chirality switching. Instead, solvent effects emerge as the decisive factor. Independent host–guest studies using a model 1,4-diethoxyP[5] host revealed that solvents such as acetonitrile and DCM exhibit substantially higher affinities for the P[5] cavity, with association constants on the order of 50–550 M^{-1}, compared with 4–18 M^{-1} for less coordinating solvents. The strong cavity affinity of polar aprotic solvents rationalizes the prevalence of out-(S_p) conformers at ambient pressure in those media, because the P[5] cavity is preferentially occupied by solvent molecules rather than by the fused subring.

Solvent-dependent stabilization of the out-conformer is further amplified for **MUJ1** and **MUJ3** which carry polyether subrings: solvation of an exposed polyether chain stabilizes the out-state and leads to more negative *g* values at ambient pressure in polar solvents than observed for **MUJ2**. Conversely, solvents with low P[5] affinity favor intramolecular self-inclusion, shifting the equilibrium toward the in-(R_p) conformer at atmospheric pressure and yielding positive *g* values.

Hydrostatic pressure perturbs this interplay via two complementary ways. First, pressurization can displace solvent molecules from the P[5] cavity, promoting subring self-inclusion and thereby increasing the fraction of in-conformers. Second, pressure can strengthen interactions between the **MUJ** portals and modestly oversized solvent molecules through a portal-perching mechanism, effectively stabilizing out-conformers in certain solvent environments. Experimental evidence suggests portal perching for solvent molecules that are too large to be fully included but can reside transiently near the ethoxy-decorated portal. Isothermal titration calorimetry and NMR chemical shift analyses on model hosts support weak but measurable interactions for THF and chloroform, suggesting that pressure-enhanced perching contributes to unusual in-to-out shifts observed in some solvents.

The combination of solvent displacement and portal perching explains why pressurization drives out-to-in transition in some solvents and in-to-out conversion in others. Polar, cavity-binding solvents, such as acetonitrile and DCM, initially populate the cavity and stabilize out-conformers. Pressurization then drives solvent exclusion and subring reentry into the cavity, causing the dramatic out-to-in transition and the observed *g* sign inversion. In less coordinating or portal-interacting solvents, perching or enhanced solvation of an exposed subring can instead stabilize the out-state when pressure increases. The net pressure effect therefore results from a delicate thermodynamic balance among intracavity guest occupancy, subring solvation, and portal interactions, each of which depends on solvent polarity, size, and shape.

The dynamic range, rapid reversibility, and reproducibility of these responses enable practical pressure-driven chiroptical switching. Repeated pressurization cycles between 40 and 200 MPa for **MUJ2** in acetonitrile produced alternating positive and negative ellipticities at comparable intensities with good reproducibility, demonstrating operational robustness as a pressure-driven chiroptical switch (Fig. 4.36). The large, solvent-tunable $\Delta g/\Delta P$ values and fully reversible sign inversion establish **MUJs** as viable building blocks for pressure-responsive molecular

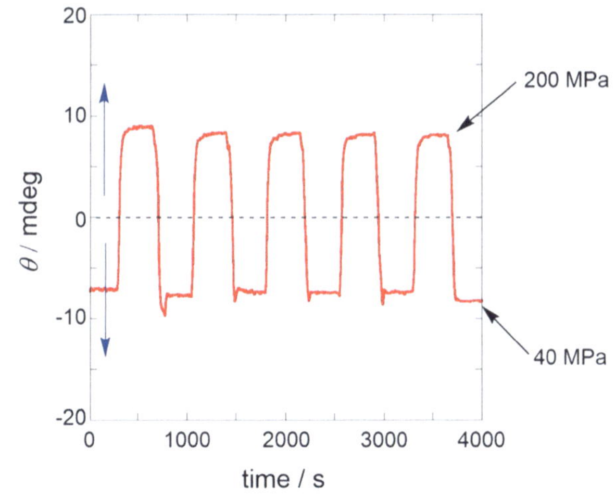

Fig. 4.36 Reversible pressure-induced switching of the ellipticity signal of in-(R_p)/out-(S_p)-**MUJ2**. Ellipticity changes at 309 nm for an acetonitrile solution of in-(R_p)/out-(S_p)-**MUJ2** (92 μM) during repeated pressurization–depressurization cycles between 40 and 200 MPa. (Reproduced with permission. Copyright 2021 Royal Society of Chemistry)

devices, sensors, and actuators that exploit solvation-directed conformational switching.

In conclusion, this study demonstrates that hydrostatic pressure serves as an orthogonal external stimulus distinct from light, thermal, redox, or chemical inputs, enabling access to conformational and functional states otherwise inaccessible. Pressure responsiveness can be programmed by tuning subring length and composition, peripheral substituent bulk, and the solvent environment, thereby controlling cavity affinity and solvation behavior of the excluded subring. Such parameters allow precise modulation of both the magnitude and direction of the pressure response, offering design strategies for piezo-sensitive materials and supramolecular machines that operate through reversible in-out switching. Spectroscopic, host-guest, and calorimetric analyses collectively reveal that pressure-induced planar chirality switching in **MUJs** originates from the interplay between subring self-inclusion and solvent inclusion, coupled with solvent-dependent stabilization of the excluded conformer and portal perching effects. These findings establish pressure as a powerful tool for the reversible control of chiroptical properties and for manipulating supramolecular assemblies through mechanisms complementary to conventional stimuli, highlighting its potential in the design of adaptive and mechanically responsive molecular systems.

4.9 Pressure-Controlled Intramolecular Singlet Fission in Pentacene Dimers

Singlet fission (SF), the process in which a single singlet exciton generates two triplet excitons, has attracted significant interest as a novel pathway for enhancing (photo)energy conversion efficiency (Smith & Michl, 2010; Miyata et al., 2019; Ullrich et al., 2021). Its appeal lies in the potential to double exciton generation from a single photon, with implications for solar cells, photodynamic therapy, optical sensors, and other advanced optoelectronic systems. Traditionally, SF has been studied under conditions defined by intrinsic molecular structures or chemical environments, but dynamic regulation through external stimuli remains challenging. This section describes how hydrostatic pressure can serve as a powerful stimulus for controlling intramolecular SF processes in solution, specifically in a biphenyl-bridged pentacene dimer (Kinoshita et al., 2023). This discovery expands the toolbox for rational SF material design and introduces a previously unexplored degree of control in excited-state photophysics.

SF is generally described as a reversible process involving chromophores in their ground state (S_0) and the excited singlet state (S_1) as shown in Fig. 4.37a. Photoexcitation generates a correlated triplet pair (TT), which subsequently separates into two individual triplet excitons (T_1). This sequence offers a quantum yield of up to 200% for triplet generation, representing an attractive mechanism compared to other modern photophysical processes. A key constraint on efficient SF is

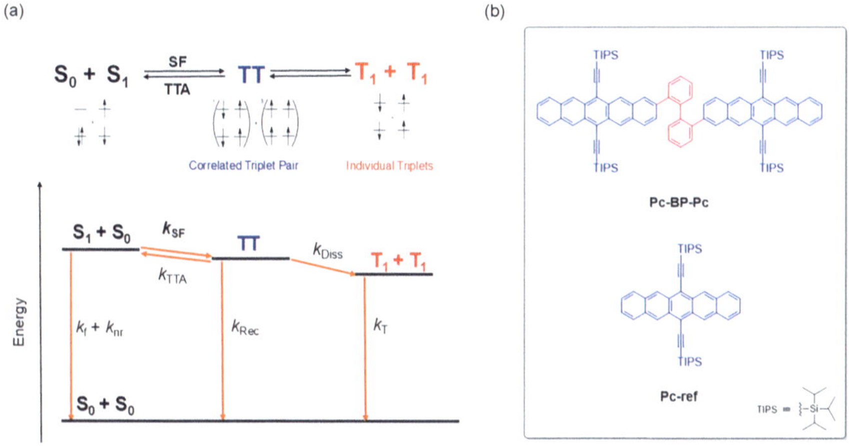

Fig. 4.37 (**a**) Schematic of the singlet fission (SF) mechanism in a TIPS-pentacene dimer. k_f and k_{nr} denote the radiative and nonradiative decay processes from S_1, k_{SF} the SF transition from S_1 to TT, k_{Rec} the recombination of TT to S_0, k_{Diss} the dissociation of TT into two triplet states ($T_1 + T_1$, $2T_1$), and k_T the deactivation of $2T_1$ to S_0. Given reported energies of S_1 (1.9 eV) and TT (1.6 eV), reverse triplet-triplet annihilation (TT → $S_0 + S_1$) is neglected due to its exergonic nature (~0.3 eV). (**b**) Chemical structures of the target compounds **Pc-BP-Pc** and **Pc-ref**. (Reproduced with permission. Copyright 2023 Royal Society of Chemistry)

energetic: the S_1 energy must be at least twice the T_1 energy. This requirement narrows the scope of molecular designs and limits flexibility in material development. External stimuli such as temperature changes, solvent polarity, aggregation, or supramolecular complexation have been tested as ways to dynamically modulate SF, but most approaches remain preliminary and lack systematic principles. Hydrostatic pressure, as shown in this work, provides an alternative strategy capable of precise control over both kinetic and thermodynamic aspects of SF.

The model system is a biphenyl-linked pentacene dimer (**Pc-BP-Pc**), whose flexible biphenyl bridge provides conformational variability distinct from its monomeric reference **Pc-ref** (Fig. 4.37b). Prior solid-state studies using diamond anvil cells demonstrated pressure-induced triplet separation in pentacene monomers, highlighting pressure as a potentially useful control variable. However, solution-state SF, particularly relevant to biomedical contexts or photodynamic therapy, had not been examined in detail. **Pc-BP-Pc** therefore provides a useful platform for probing how pressure and solvent environments jointly influence intramolecular SF dynamics in a dispersed, monomeric state.

UV-vis absorption measurements confirmed that **Pc-BP-Pc** remained well-dispersed under pressures up to 320 MPa, with no evidence of pressure-induced aggregation or crystallization (Fig. 4.38). This supports assignment of the observed pressure effects to genuine molecular-level processes rather than macroscopic aggregation. Spectroscopic analysis of both **Pc-BP-Pc** and **Pc-ref** as a monomer reference showed monotonic bathochromic shifts in absorption and fluorescence spectra under pressure, consistent with known solvent polarizability effects.

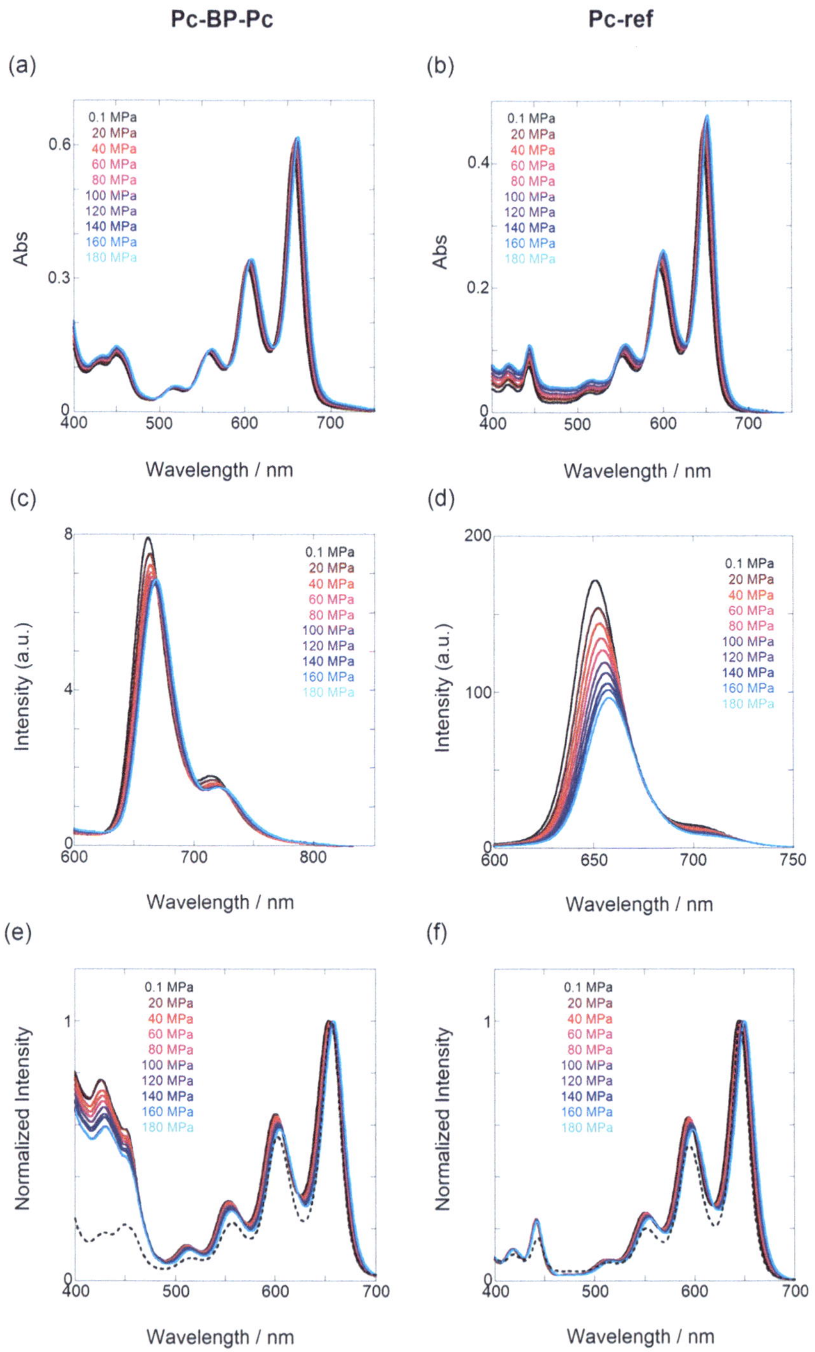
Pc-BP-Pc
Pc-ref
(a)
(b)
(c)
(d)
(e)
(f)
0.1 MPa
20 MPa
40 MPa
60 MPa
80 MPa
100 MPa
120 MPa
140 MPa
160 MPa
180 MPa
Abs
Intensity (a.u.)
Normalized Intensity
Wavelength / nm

Importantly, no evidence was found for intramolecular π-stacking in either ground or excited states, reinforcing the notion that pressure-induced changes stemmed from solvation and conformational effects rather than aggregation.

Fluorescence lifetime measurements revealed critical insights into the SF dynamics. For **Pc-BP-Pc**, fluorescence decays were biexponential, with long-lived components corresponding to monomer-like fluorescence and short-lived components reflecting the SF pathway (Fig. 4.39a). Conversely, **Pc-ref** showed simple monoexponential decay, consistent with the absence of SF (Fig. 4.39b). Under hydrostatic pressure, the apparent SF rate constant ($k_{SF,app}$) for **Pc-BP-Pc** increased systematically (Fig. 4.39c). In toluene, $k_{SF,app}$ rose from $1.65 \times 10^9\ s^{-1}$ at atmospheric pressure to $1.96 \times 10^9\ s^{-1}$ at 180 MPa, while in THF a similar increasing trend was observed. In contrast, in nonpolar MCH, the SF rate remained essentially unchanged. This contrast highlights the decisive role of solvent polarity and solvation dynamics in mediating pressure effects on SF.

Quantitative analysis of activation volumes provided deeper mechanistic insight. The activation volume change ($\Delta V^‡$), derived from the pressure-dependent rate constants, was negative in toluene and THF but approximately zero in MCH. A negative $\Delta V^‡$ indicates a more compact transition state for SF under pressure in polar solvents due to desolvation (Fig. 4.39d). The compactness of the TT intermediate likely reflects the expulsion or reorganization of solvent molecules, which facilitates efficient triplet pair formation. This desolvation-driven compaction accelerates SF by lowering the transition-state volume; however, this does not occur in nonpolar solvents where solvation effects are weaker. Thus, solvation and desolvation emerge as critical modulators of SF efficiency under pressure.

nsTA measurements were used to probe the triplet state (T_1) dynamics following SF (Fig. 4.40a). Under pressure, the nsTA spectra revealed modest bathochromic shifts, while triplet lifetimes decreased significantly with increasing pressure, from microsecond timescales toward shorter values (Fig. 4.40b, c). This lifetime shortening was not attributed to the intrinsic instability but rather to the enhanced solvent interactions in higher viscosity conditions induced by pressure. These results suggest that while hydrostatic pressure accelerates SF, it also increases deactivation of individual triplets due to enhanced solvent coupling, pointing to a trade-off between kinetic acceleration and exciton persistence.

Thermodynamic analysis further clarified the overall process (Fig. 4.40d). The equilibrium volume change for the $S_1 \rightarrow TT$ step was negative, consistent with a compact TT state stabilized under pressurization via desolvation. Conversely, the

Fig. 4.38 (**a**–**f**) Pressure-dependent spectral changes of (**a**, **c**, **e**) **Pc-BP-Pc** and (**b**, **d**, **f**) **Pc-ref** in toluene at room temperature. (**a**, **b**) UV-vis absorption spectra ((**a**) 82 μM, (**b**) 91 μM); (**c**, **d**) fluorescence spectra ((**c**) 67 μM, λ_{ex} = 580 nm; (**d**) 91 μM, λ_{ex} = 543 nm); and (**e**, **f**) excitation spectra ((**e**) 82 μM, λ_{em} = 660 nm; (**f**) 91 μM, λ_{em} = 650 nm) recorded under 0.1–180 MPa (0.1, 20, 40, 60, 80, 100, 120, 140, 160, and 180 MPa, from black to sky blue) using a high-pressure cell. Black dotted lines indicate the normalized UV-vis spectra at 0.1 MPa. (Reproduced with permission. Copyright 2023 Royal Society of Chemistry)

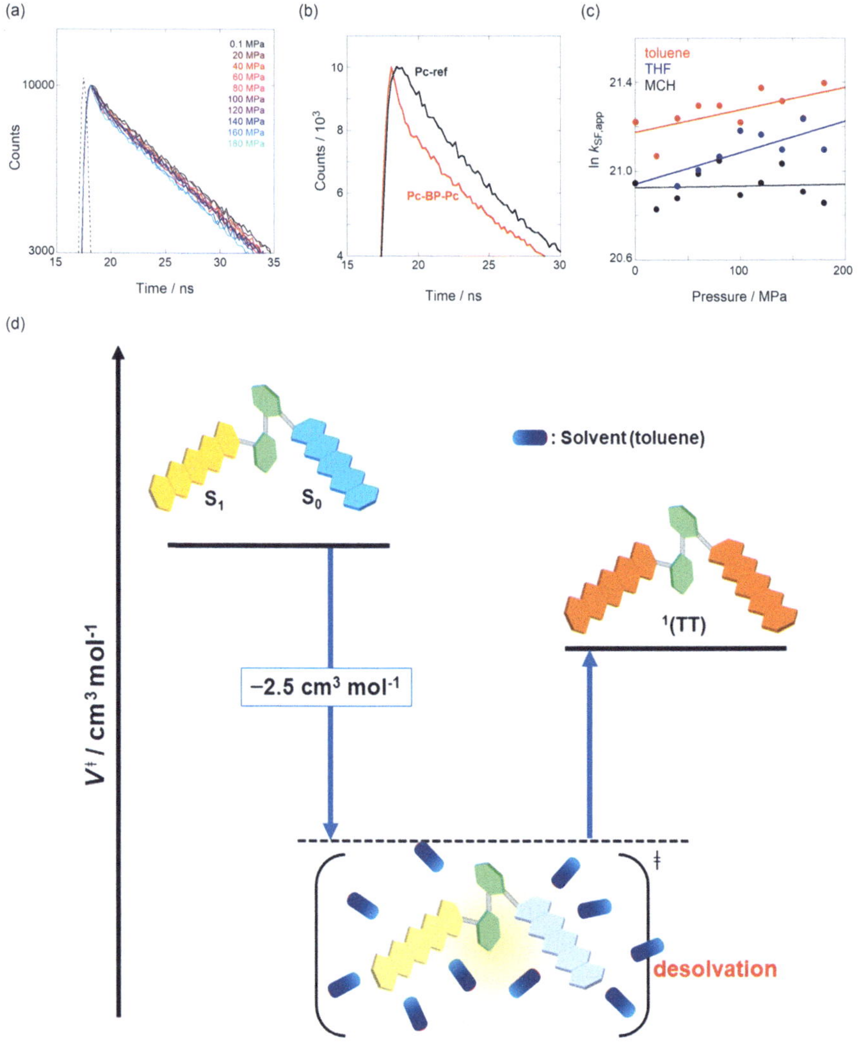

Fig. 4.39 (**a**, **b**) Enlarged time-correlated fluorescence decay profiles (λ_{ex} = 405 nm) of (**a**) **Pc-BP-Pc** (74 µM, λ_{em} = 660 nm) in toluene at room temperature under 0.1–180 MPa (0.1, 20, 40, 60, 80, 100, 120, 140, 160, and 180 MPa; from black to sky blue), and (**b**) **Pc-BP-Pc** (74 µM, λ_{em} = 660 nm, red) and **Pc-ref** (89 µM, λ_{em} = 650 nm, black) in toluene at 180 MPa, measured using a high-pressure cell. The black dotted line indicates the instrument response function. (**c**) Pressure dependence of the apparent SF rate constant ($k_{SF,app}$) of **Pc-BP-Pc** in toluene (red, r = 0.655), THF (blue, r = 0.706), and MCH (black, r = 0.059) at room temperature. The apparent deviations in r reflect the limited precision of the lifetime apparatus. (**d**) Schematic of **Pc-BP-Pc** illustrating volumetric changes in the transition state during the intramolecular SF process. (Reproduced with permission. Copyright 2023 Royal Society of Chemistry)

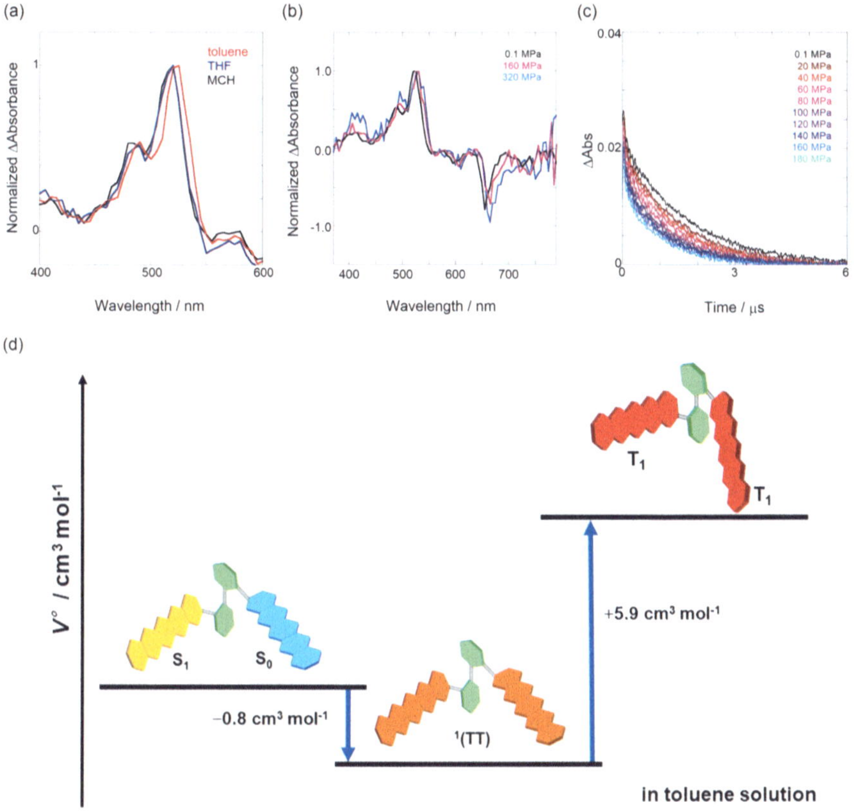

Fig. 4.40 (**a**, **b**) Normalized nsTA spectra of **Pc-BP-Pc** (81 μM): (**a**) in toluene (red), THF (blue), and MCH (black) at 0.1 MPa, and (**b**) in toluene at 0.1 MPa (black), 160 MPa (pink; the line between 510 and 520 nm was omitted due to an artifact), and 320 MPa (blue) at room temperature. (**c**) nsTA decay profiles of **Pc-BP-Pc** in toluene (81 μM, $\lambda_{ex} = 532$ nm, $\lambda_{obs} = 521$ nm) at 0.1–180 MPa (0.1, 20, 40, 60, 80, 100, 120, 140, 160, and 180 MPa; from black to sky blue). (**d**) Schematic of the volumetric changes in the equilibria associated with intramolecular SF and dissociation processes of **Pc-BP-Pc**. (Reproduced with permission. Copyright 2023 Royal Society of Chemistry)

$TT \rightarrow 2T_1$ dissociation step exhibited positive reaction volumes, suggesting conformational reorganization of the dimer during dissociation. This implies that TT dissociation requires structural relaxation, likely involving torsional motion around the biphenyl linker, leading to larger effective molecular volumes. These volumetric trends explain the observed decrease in triplet quantum yields under pressure: as TT dissociation becomes associated with bulkier conformations and stronger solvent interactions, the deactivation pathways compete more effectively, reducing exciton lifetimes.

In conclusion, this study establishes hydrostatic pressure as a precise and versatile stimulus for modulating intramolecular SF. Pressure accelerates SF kinetics through desolvation-driven transition-state compaction, while also influencing dissociation pathways and triplet lifetimes via conformational and viscosity effects. Taken together, these results reveal that hydrostatic pressure enables precise regulation of both SF ($S_0 + S_1 \rightarrow TT$) and dissociation ($TT \rightarrow 2T_1$) through a combination of kinetic transition-state effects and thermodynamic equilibrium shifts driven by desolvation and solvent reorientation. These findings not only complement traditional approaches based on structural or chemical modification but also establish solvation and desolvation as decisive factors, offering a framework for designing next-generation SF materials responsive to mechanical stimuli. Such insights are expected to guide future advances in solar energy harvesting, photodynamic therapy, and molecular optoelectronics.

4.10 Molecular Design for Active Pressure Control of Intramolecular Singlet Fission

Building on the hydrostatic pressure-controlled intramolecular SF established for the biphenyl-linked pentacene dimer **Pc-BP-Pc**, this section examines whether similar volumetric and solvation principles operate in a wider class of soft pentacene dimers with more flexible linkers. Specifically, pentacene dimers that incorporate ethane and cyclohexane-based linkers, together with the corresponding monomeric references, are compared with the biphenyl-bridged system to clarify how linker flexibility and pressure-dependent solvation govern intramolecular SF and the subsequent formation and decay of individual triplet excitons (Fig. 4.41) (Ogawa et al., 2025).

The role of linker rotation in pressure-responsive SF was first evaluated by DFT calculations on model linkers **BP**, **Cy**, and **Et**. Potential energy profiles as a function of dihedral angle reveal that the biphenyl unit has the highest rotational barriers (~69 and 88 kJ mol^{-1}), the cyclohexane linker has a lower barrier (~55 kJ mol^{-1}), and the ethane chain shows a barrier of 28 kJ mol^{-1}, which is less than half that for biphenyl (Fig. 4.42). The order of rotational barrier **BP** > **Cy** > **Et** indicates that the new dimers with **Cy** and **Et** linkers should exhibit stronger responses to pressure-induced volume changes than the former biphenyl-based system.

Before considering excited-state processes, pressure experiments were used to confirm that the dimers behave as isolated molecules in solution. UV-vis spectra of **Pc-EE-Pc** in toluene up to 160 MPa showed normalized absorption profiles that are identical over a wide concentration range, and absorbance varied linearly with concentration at both ambient and elevated pressure (Fig. 4.43). These results rule out aggregation and crystallization and justify treating SF as an intramolecular process. Similar behavior is observed for the other dimers, with one important exception: **Pc-CA-Pc** in MCH showed a slight red shift of the absorption maximum with increasing concentration, consistent with *J*-type aggregation mediated by hydrogen

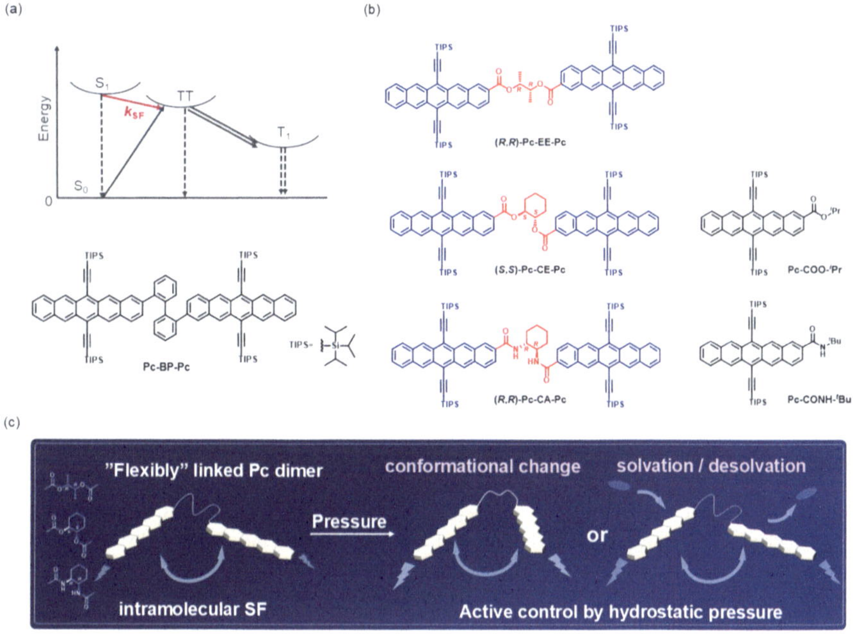

Fig. 4.41 (**a**) (Top) Schematic of SF in a TIPS-substituted pentacene (Pc) dimer. The rate constant k_{SF} denotes the transformation from S_1 to the correlated TT state. (Bottom) Chemical structure of **Pc-BP-Pc**. (**b**) Chemical structures of the SF-active dimers, **(*R*,*R*)-Pc-EE-Pc**, **(*S*,*S*)-Pc-CE-Pc**, and **(*R*,*R*)-Pc-CA-Pc**, together with their respective reference monomers, **Pc-COO-^{i}Pr** and **Pc-CONH-^{t}Bu**. (**c**) Conceptual illustration depicting the active modulation of SF-active soft materials under hydrostatic pressure. (Reproduced with permission. Copyright 2025 Royal Society of Chemistry)

bonding at the amide. Accordingly, **Pc-CA-Pc** is discussed under pressure only in toluene, whereas **Pc-EE-Pc** and **Pc-CE-Pc** are examined in both toluene and MCH.

Hydrostatic pressure-dependent UV-vis absorption and CD measurements clarified the ground-state response. In toluene, all three dimers and both monomers displayed monotonic hyperchromic changes and small bathochromic shifts in absorption upon increasing pressure. The intensity increase reflects the enhanced effective concentration caused by pressurization, and the red shift is attributed to increased solvent density and polarizability. Depressurization restored the original spectra, demonstrating reversibility. The magnitude of the red shift is similar for dimers and monomers and comparable to values reported for other aromatic systems, which supports the conclusion that no specific ground-state conformational change of the dimer skeleton occurs under pressure. CD and *g* factor spectra of the chiral dimers showed clear bisignate exciton couplets whose sign identifies the handedness of the exciton coupling (Fig. 4.44). Crucially, the amplitude of the *g* factor remained essentially constant as pressure increased, despite small red shifts in the underlying absorption, again indicating that the relative arrangement of the pentacene units and the conformations around the flexible linkers do not change

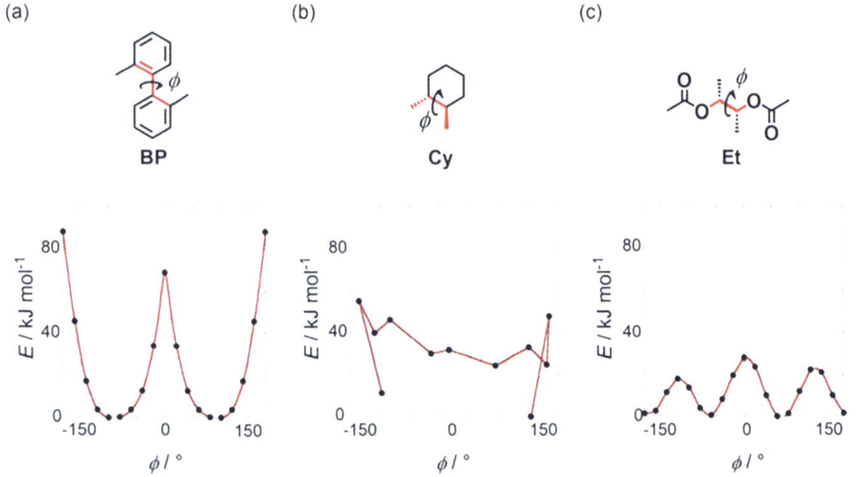

Fig. 4.42 Potential energy profiles as a function of the dihedral angle for (**a**) **BP**, (**b**) **Cy**, and (**c**) **Et**, calculated at the w97XD/6-31G(d,p) level. For each linker, the most stable conformer is defined as 0 kJ mol^{-1}. The abrupt hysteresis-like discontinuity observed in (**b**) arises from a ring-flip event. (Reproduced with permission. Copyright 2025 Royal Society of Chemistry)

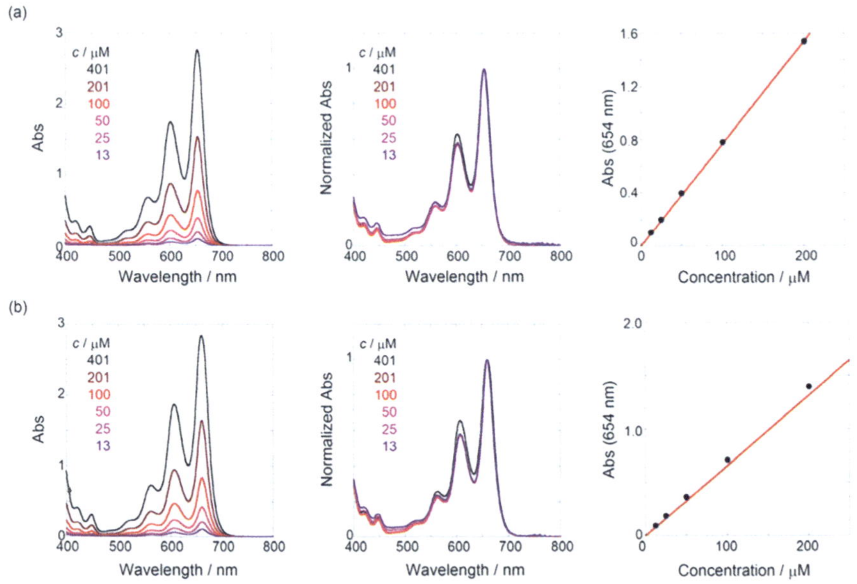

Fig. 4.43 Concentration-dependent UV-vis spectra of **Pc-EE-Pc** in toluene (13–401 μM, left), corresponding normalized spectra (middle), and absorbance-concentration plots (13–201 μM, right), recorded at room temperature using a high-pressure cell. Measurements were performed at (**a**) 0.1 MPa, yielding a correlation coefficient of $r = 1.000$ and a slope of 0.0077 (±3.9 × 10^{-1}) μM $^{-1}$, and at (**b**) 160 MPa with $r = 0.999$ and a slope of 0.0064 (±0.00013) μM $^{-1}$. (Reproduced with permission. Copyright 2025 Royal Society of Chemistry)

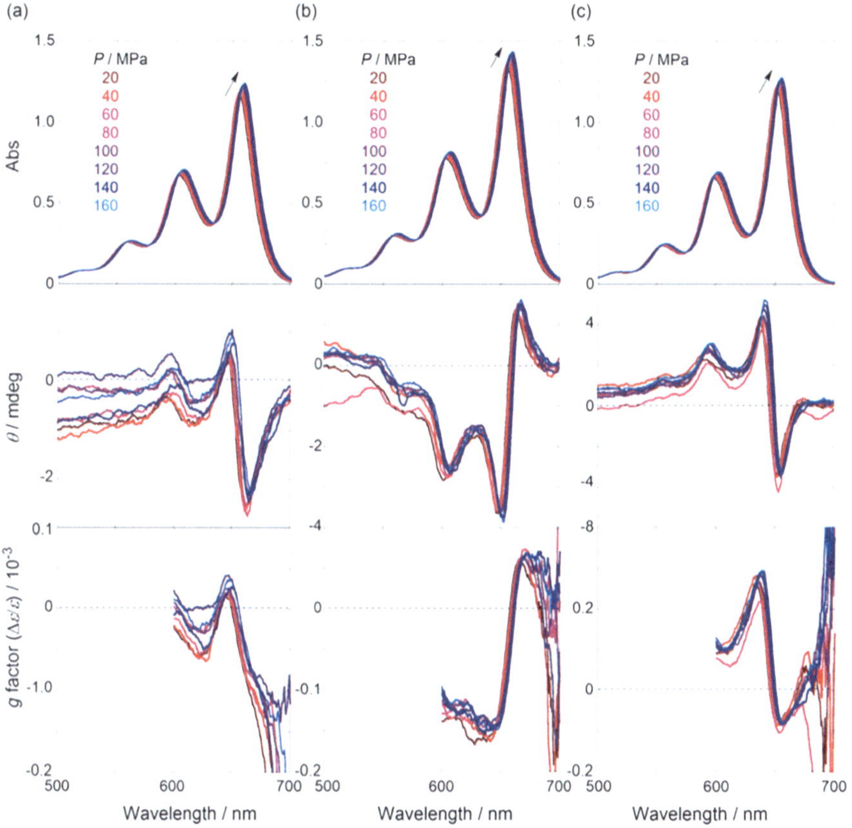

Fig. 4.44 Pressure-dependent UV-vis absorption (top), CD (middle), and g factor spectra ($g = \Delta\varepsilon/\varepsilon$, bottom) of (**a**) **(*R*,*R*)-Pc-EE-Pc** (144 μM), (**b**) **(*S*,*S*)-Pc-CE-Pc** (142 μM), and (**c**) **(*R*,*R*)-Pc-CA-Pc** (136 μM) in toluene at room temperature recorded in a high-pressure cell at 20, 40, 60, 80, 100, 120, 140, and 160 MPa. The g factor traces below 600 nm were excluded due to baseline noise. (Reproduced with permission. Copyright 2025 Royal Society of Chemistry)

significantly in the ground state. These observations establish that the pressure responsive behavior of SF arises from excited-state kinetics rather than pressure-driven ground-state structural changes.

Intramolecular SF at ambient pressure was quantified by fluorescence lifetime measurements in toluene (Fig. 4.45). The monomeric references showed single exponential decays with lifetimes of ~11–13 ns, whereas all three dimers exhibit biexponential decays with shorter components. The longer component is assigned to emissions from locally excited pentacene sites that do not undergo SF, while the shorter component reflects population that decays via the SF channel. A simple kinetic expression (Eq. 4.1) relates an apparent SF rate constant $k_{\mathrm{SF,app}}$ to the monomer lifetime ($\tau_{1,\mathrm{ref}}$) and the short dimer lifetime ($\tau_{1,\mathrm{dimer}}$).

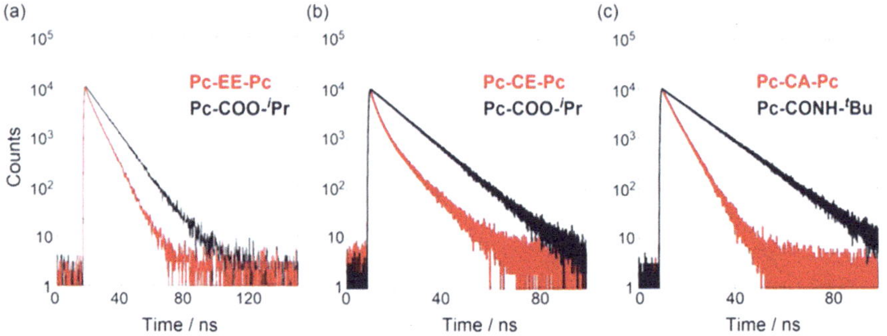

Fig. 4.45 Time-correlated fluorescence decay profiles (λ_{ex} = 590 nm) of (**a**) **Pc-EE-Pc** (44 μM, λ_{obs} = 665 nm, red) compared with **Pc-COO-^{i}Pr** (82 μM, λ_{obs} = 665 nm, black), (**b**) **Pc-CE-Pc** (41 μM, λ_{obs} = 665 nm, red) compared with **Pc-COO-^{i}Pr** (82 μM, λ_{obs} = 665 nm, black), and (**c**) **Pc-CA-Pc** (41 μM, λ_{obs} = 660 nm, red) compared with **Pc-CONH-^{t}Bu** (89 μM, λ_{obs} = 656 nm, black), recorded in toluene at room temperature under 0.1 MPa using a high-pressure cell. (Reproduced with permission. Copyright 2025 Royal Society of Chemistry)

$$k_{\mathrm{SF,app}} = \left(k_{\mathrm{SF}} + k_0\right) - k_0 = \frac{1}{\tau_{1,\mathrm{dimer}}} - \frac{1}{\tau_{1,\mathrm{ref}}} \tag{4.1}$$

where k_{SF} and k_0 represent the rate constant of SF and the decay rate constant from the singlet excited state of the reference monomer. Application of this relation yielded $k_{SF,app}$ values of 6.84 × 10^8 s^{-1} for **Pc-EE-Pc**, 3.33 × 10^8 s^{-1} for **Pc-CE-Pc**, and 2.54 × 10^8 s^{-1} for **Pc-CA-Pc** in toluene at 0.1 MPa.

Picosecond transient absorption spectra (psTA) of **Pc-CE-Pc** in degassed toluene show immediate formation of a singlet excited state absorption band between 450 and 600 nm, followed by its decay and growth of a triplet absorption band between 450 and 550 nm (Fig. 4.46). The decay at a singlet selective wavelength yields a lifetime of 3.4 ns, while the triplet signal shows a rise with the same time constant and a decay ~21 μs. The reciprocal of the singlet decay time gives a rate near 3.0 × 10^8 s^{-1}, in close agreement with $k_{SF,app}$ extracted from fluorescence, which confirms the reliability of the lifetime-based approach.

Hydrostatic pressure-dependent fluorescence decays in toluene permit extraction of $k_{SF,app}$ as a function of pressure (Fig. 4.47). For each dimer, the biexponential character persists under pressure, while the monomer reference remains monoexponential and shows a gradual lifetime increase with pressurization (Fig. 4.47a). Plots of ln$k_{SF,app}$ versus pressure are linear, with their slopes corresponding to activation volume changes ($\Delta V^{\ddagger}$) for formation of the correlated triplet pair (Fig. 4.47b). The biphenyl-linked dimer exhibits a negative $\Delta V^{\ddagger}$ (−2.5 cm^3 mol^{-1}), consistent with transition state compaction via desolvation of a relatively non polar linker. The flexible dimers show positive $\Delta V^{\ddagger}$ of 3.0 cm^3 mol^{-1} for **Pc-EE-Pc**, 2.5 cm^3 mol^{-1} for **Pc-CE-Pc**, and 2.7 cm^3 mol^{-1} for **Pc-CA-Pc**. These positive values are attributed to spontaneous solvation around polar carbonyl or amide groups in the transition state,

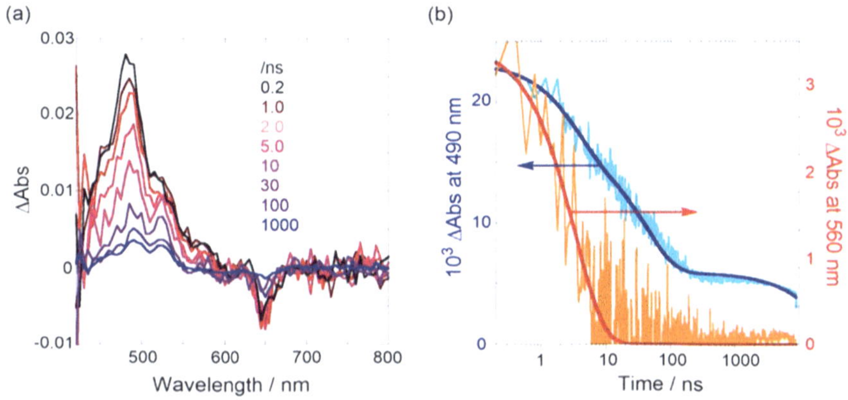

Fig. 4.46 (**a**) psTA spectra and (**b**) the corresponding temporal profile of **Pc-CE-Pc** (64 μM) recorded in degassed toluene at room temperature under 0.1 MPa. The psTA measurements were performed at 532 nm excitation in a 2 mm path-length cell. (Reproduced with permission. Copyright 2025 Royal Society of Chemistry)

which expands the total volume (Fig. 4.47c). Comparable values for the three flexible dimers support a common mechanism in which solvation and desolvation around polar linkers modulate SF.

Solvent-dependent experiments extend this interpretation. Activation volumes for SF in MCH, toluene, and DCM were determined for **Pc-EE-Pc** and **Pc-CE-Pc** (Fig. 4.48a). In the less polar solvents, MCH and toluene, SF slows under pressure and $\Delta V^{\ddagger}$ remains positive. In DCM, SF in **Pc-CE-Pc** becomes slightly faster as pressure increases and $\Delta V^{\ddagger}$ becomes negative. A compensation plot of $\Delta V^{\ddagger}$ versus the Reichardt E_T (30) polarity parameter shows an almost linear decrease as polarity increases from MCH to toluene to DCM (Fig. 4.48b). This trend is interpreted in terms of solvation of the excited carbonyl group: in weakly solvating media, increased pressure enhances solvation around polar sites and expands the transition-state volume, while in a highly solvating medium such as DCM, additional solvation becomes unfavorable and desolvation of the solvent molecules leads to compaction and negative $\Delta V^{\ddagger}$ (Fig. 4.48c). Thus, the hydrostatic pressure and solvent polarity cooperate to shift SF between regimes dominated by solvation and desolvation.

Triplet exciton dynamics after intramolecular SF were examined by nsTA under pressure (Fig. 4.49). In toluene, **Pc-EE-Pc** shows a strong excited-state absorption band assigned to transitions from T_1 to higher triplet states between 400 and 550 nm (Fig. 4.49a). Kinetic traces at a representative wavelength reveal monoexponential decays whose lifetimes decrease from ~12.2 μs at ambient pressure to 7.5 μs at 160 MPa (Fig. 4.49b). Similar pressure-induced lifetime shortening is observed for the other dimers and in MCH, consistent with the increased solvent viscosity under pressure. Such a finding demonstrates that triplet deactivation can be tuned by hydrostatic compression without altering the molecular structure.

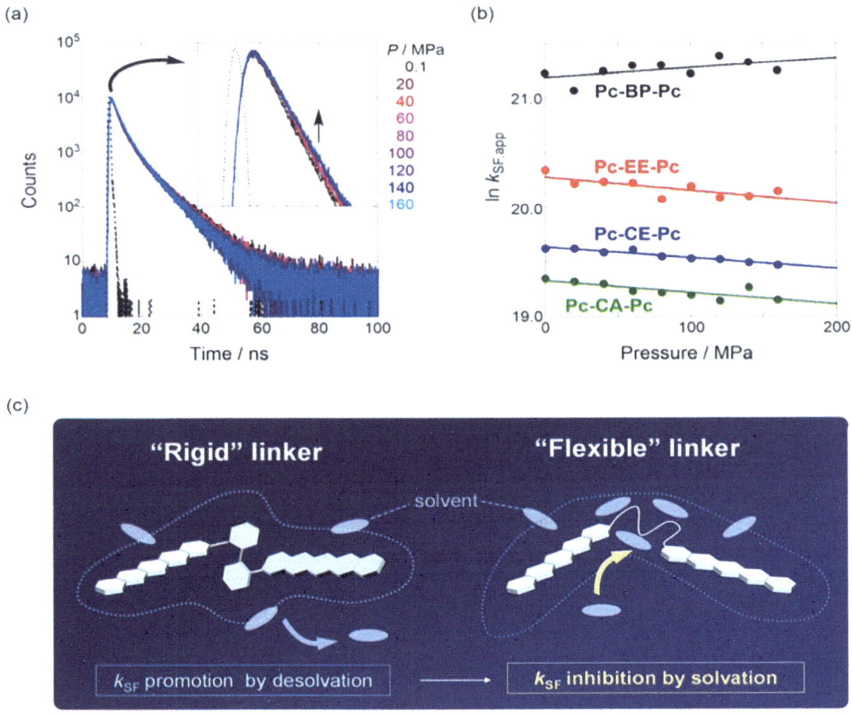

Fig. 4.47 (**a**) Pressure-dependent time-correlated fluorescence decay traces (λ_{ex} = 590 nm, λ_{obs} = 665 nm) of **Pc-CE-Pc** (41 μM) in toluene, recorded at 0.1, 20, 40, 60, 80, 100, 120, 140, and 160 MPa (from black to sky-blue). Measurements were performed at room temperature using a high-pressure cell. The inset shows an expanded view of the decay curves, and the black dotted line denotes the instrument response function. (**b**) Pressure dependence of the apparent SF rate constants ($k_{SF,app}$) for **Pc-BP-Pc** (black, slope = 0.00089 ± 0.00051), **Pc-EE-Pc** (red, slope = −0.0012 ± 0.00038), **Pc-CE-Pc** (blue, slope = −0.0010 ± 0.00011), and **Pc-CA-Pc** (green, slope = −0.0011 ± 0.00029) in toluene at room temperature. (**c**) Conceptual illustration comparing the total molecular volumes of the Pc dimers equipped with "Rigid" versus "Flexible" linkers. (Reproduced with permission. Copyright 2025 Royal Society of Chemistry)

The equilibria between the correlated triplet pairs and independent triplets were analyzed by quantifying triplet pair and individual triplet quantum yields (Φ_{TT} and Φ_{T}, respectively) and applying a standard relation between equilibrium constants and reaction volumes (Fig. 4.50). In toluene, the individual triplet quantum yields of the flexible dimers remain nearly constant as pressure increases, in contrast to the marked decrease observed previously for the biphenyl-linked system (Fig. 4.50a, b). Reaction volume changes ($\Delta V^{\circ}_{TT \to T}$) for conversion from triplet pair to two triplets are small and negative for **Pc-EE-Pc**, positive for **Pc-CE-Pc** with a larger magnitude in MCH, and slightly negative for **Pc-CA-Pc** (Fig. 4.50c). These values are rationalized by differences in linker congestion: an open ethane linker favors

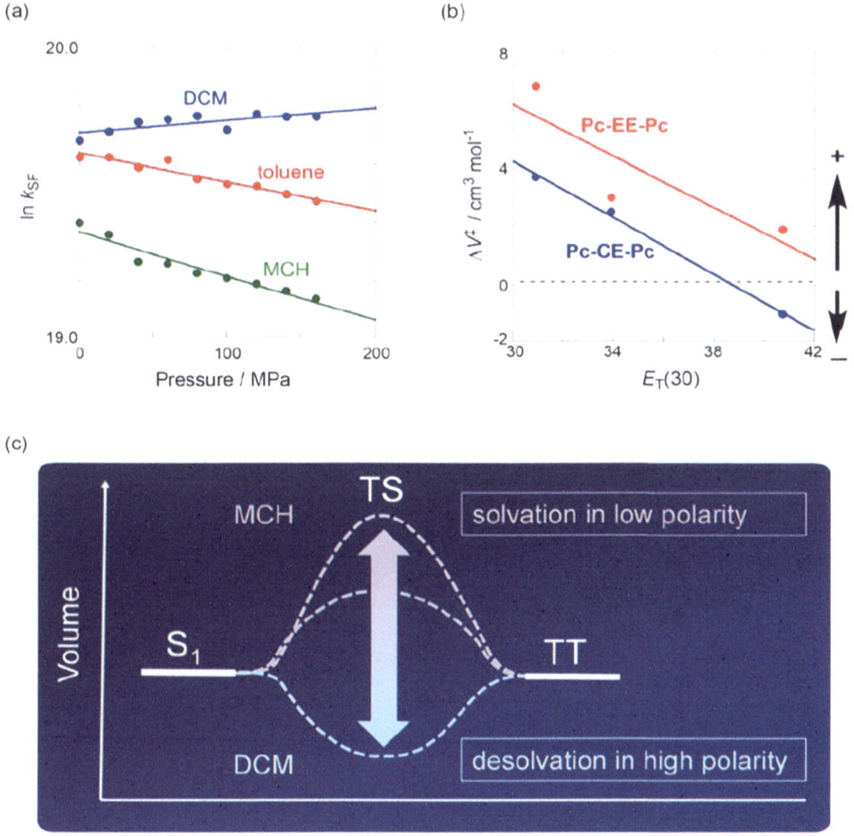

Fig. 4.48 (**a**) Pressure dependence of the apparent SF rate constants ($k_{SF,app}$) for **Pc-CE-Pc** measured in DCM (blue, $r = 0.733$, slope = 0.00042 ± 0.00015), toluene (red, $r = 0.963$, slope = -0.0010 ± 0.00011), and MCH (green, $r = 0.965$, slope = -0.0015 ± 0.00016). (**b**) Compensation plot of $\Delta V^{\ddagger}$ versus $E_T(30)$ for **Pc-EE-Pc** (red, $r = 0.864$, slope = -0.44 ± 0.26) and **Pc-CE-Pc** (blue, $r = 0.998$, slope = -0.49 ± 0.027). (**c**) Schematic volumetric representation of the iSF process progressing toward the TT state. (Reproduced with permission. Copyright 2025 Royal Society of Chemistry)

desolvation and volume contraction when two triplets separate, whereas a more congested cyclohexane linker can trap solvent and cause volume expansion. Such results show that flexible and polar linkers enable active switching between faster and slower SF, allow shortening of triplet lifetimes without loss of triplet yield, and permit control over triplet formation thermodynamics under hydrostatic pressure. This critical molecular design strategy provides a basis for pressure responsive soft materials that exploit SF in photophysical and photochemical applications.

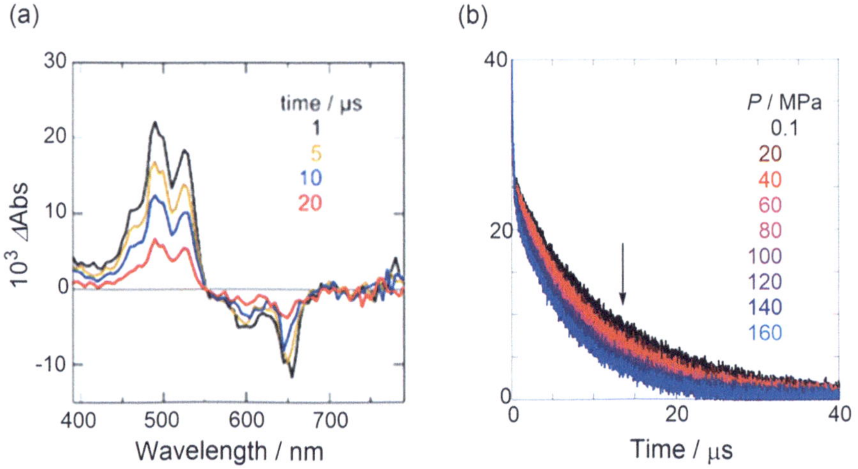

Fig. 4.49 (**a**) nsTA spectra of **Pc-EE-Pc** at 0.1 MPa and (**b**) corresponding decay profiles monitored at 495 nm in degassed toluene at 0.1, 20, 40, 60, 80, 100, 120, 140, and 160 MPa (from black to sky blue lines) at room temperature, measured in a high-pressure cell. (Reproduced with permission. Copyright 2025 Royal Society of Chemistry)

In summary, hydrostatic pressure serves as an effective external stimulus for actively modulating intramolecular singlet fission in soft molecular systems when chromophores are connected through linkers of appropriate flexibility and polarity. Systematic comparisons across linkers with different rotational freedoms clarify that conformational softness, combined with pressure-dependent solvation and desolvation around polar sites, governs not only the rate at which correlated triplet pairs are generated but also the thermodynamics of their separation into individual triplet excitons. Pressure affects excited-state pathways rather than ground-state structures, enabling reversible switching between accelerated and decelerated singlet fission, tuning of triplet lifetimes through solvent viscosity, and preservation of triplet quantum yields even under substantial compression. These findings establish a molecular design principle in which linker flexibility and local polarity dictate the volumetric response of excited state processes. Such a strategy provides a foundation for creating soft materials whose SF behavior and triplet exciton properties can be programmed by hydrostatic pressure, opening opportunities for pressure responsive photochemical functions and advanced photosensitizing applications.

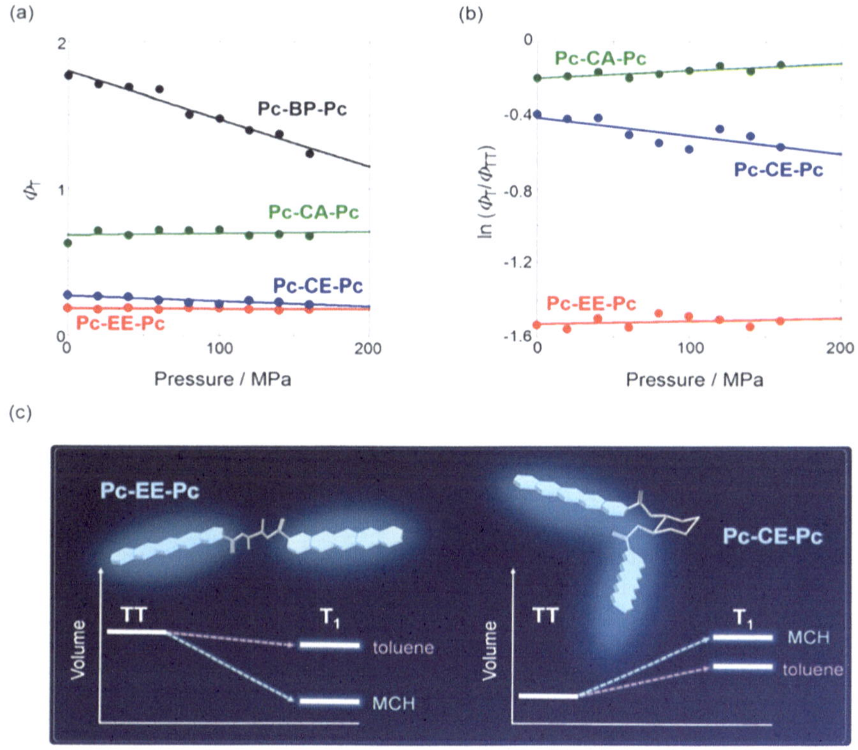

Fig. 4.50 Pressure dependence of (**a**) T_1 quantum yields (Φ_T) and (**b**) relative abundance ratios (Φ_T/Φ_{TT}) for **Pc-BP-Pc** ($r = 0.979$, slope $= -0.0032 \pm 0.00025$), **Pc-EE-Pc** (red, $r = 0.222$, slope $= -2.1 \times 10^{-5} \pm 3.5 \times 10^{-5}$ for (**a**), and $r = 0.287$, slope $= 0.00016 \pm 0.00020$ for (**b**)), **Pc-CE-Pc** (blue, $r = 0.861$, slope $= -0.00035 \pm 7.8 \times 10^{-5}$ for (**a**), and $r = 0.764$, slope $= -0.00099 \pm 0.00032$ for (**b**)), and **Pc-CA-Pc** (green, $r = 0.240$, slope $= 0.00013 \pm 0.00020$ for (**a**), and $r = 0.821$, slope $= 0.00041 \pm 0.00011$ for (**b**)) in toluene at room temperature. (**c**) Volumetric representation illustrating the distinct pathways for T_1 exciton formation in the two Pc dimers. (Reproduced with permission. Copyright 2025 Royal Society of Chemistry)

4.11 Redox Control Through Hydrostatic Pressure in Solvated Cyclophanes

Hydrostatic pressure is treated in this section as a mechanical, isotropic stimulus that can shift equilibria and reaction rates in solution by acting through the reaction-volume term. Conventional discussions on solution thermodynamics focus on ΔG expressed in terms of ΔH and ΔS at a constant pressure, where the ΔS contribution is dominated by complex solvation effects and is hard to design in a predictive way. Under constant temperature, ΔG instead separates into ΔF and $P\Delta V$. If the applied pressure is sufficiently large and ΔV is negative, the overall driving force becomes favorable. Since the volume term reflects structural changes, solvation and desolvation, and changes in the van der Waals volume, it is more amenable to rational control than the entropy term and provides an attractive handle for pressure-responsive chemistry.

Cyclophanes provide a particularly promising platform in this context because the inclusion and release of guests, and the associated conformational flexibility, involve clear changes in the van der Waals volume. Cationic cyclophanes such as Stoddart-type systems are already known as redox active hosts whose properties can be tuned by electrical potential, but pressure-controlled redox chemistry has remained largely unexplored, in part because simple organic redox couples often show only small volume changes upon electron transfer. The dicationic cyclophane $\mathbf{2}^{2+}$ is identified as a candidate in which the redox state should strongly influence the molecular structure (Fig. 4.51) (Kikuchi et al., 2025). This macrocycle contains two xanthylium units linked by flexible alkylene spacers, so both the distance between the π skeletons and the overall conformation can change in solution, unlike a monomeric reference $\mathbf{1}^{+}$. Decomposition of the apparent reaction volume into intramolecular and solvation contributions shows that a one-electron reduction from $\mathbf{2}^{2+}$ to a radical cation state $\mathbf{2}^{\bullet+}$ is expected to decrease both the internal van der Waals volume, through closer π stacking as electrostatic repulsion is reduced, and the solvated molar volume, through desolvation that accompanies the lower net charge. A strongly negative reaction volume is therefore anticipated and should make the reduction more favorable under hydrostatic pressure.

The system was constructed so that redox processes could be monitored and quantified spectroscopically. The cyclophane dication $\mathbf{2}^{2+}$ undergoes a reversible two-step, one-electron reduction sequence that gives a radical cation $\mathbf{2}^{\bullet+}$ and then an isolable neutral biradical $\mathbf{2}^{2\bullet}$. Although $\mathbf{2}^{\bullet+}$ cannot be isolated, the UV-vis spectra of $\mathbf{2}^{2+}$ and $\mathbf{2}^{\bullet+}$ can be distinguished, and molar extinction coefficients for the radical cation can be obtained by measuring equimolar mixtures of $\mathbf{2}^{2+}$ and the neutral biradical. Substituents at the xanthylium 9 position are varied between 4-methoxyphenyl and 5-(4-methoxyphenyl)thienyl groups to probe the influence of π conjugation. Three different counter anions, BF_4^-, PF_6^-, and NTf_2^-, were selected to explore the effect of anion polarization and solvation on redox behavior, and salts of $\mathbf{2}^{2+}$ with PF_6^- and NTf_2^- were prepared analogously to the known BF_4^- salt.

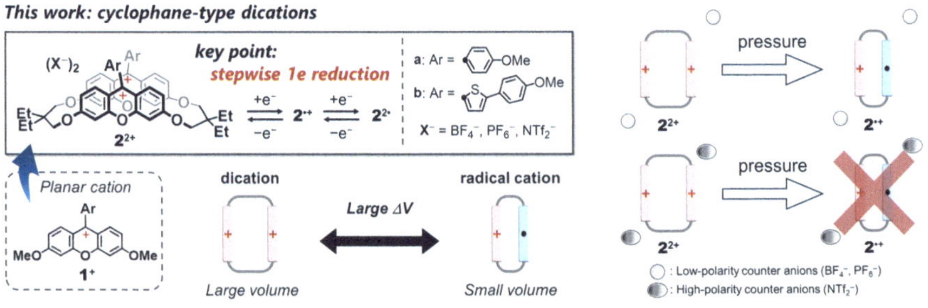

Fig. 4.51 Schematic of the system presented in this section. (Reproduced with permission. Copyright 2025 Royal Society of Chemistry)

Under ambient pressure in dry DCM, UV-vis spectra of the three dicationic salts show intense visible absorption bands assigned to the xanthylium chromophores. The absorption maxima and molar extinction coefficients are very similar for BF_4^-, PF_6^-, and NTf_2^- for each cationic framework, which indicates that ion pairing is negligible in these dilute solutions and that counter anion identity does not significantly perturb the electronic structure of $\mathbf{2}^{2+}$ in the ground state.

The response of $\mathbf{2}^{2+}$ to hydrostatic pressure was then examined by pressure-dependent UV-vis spectroscopy (Fig. 4.52). In dehydrated DCM, the spectra of the BF_4^- salt show a gradual increase in absorbance and a slight red shift with increasing pressure, without qualitative changes in band shape. This behavior corresponds to the typical pressure response of π-conjugated molecules and indicates that the dicationic state is preserved under pressurization in the absence of water (Fig. 4.52b). In contrast, in a DCM/water mixture containing several thousand equivalents of water per cyclophane, the spectral changes are qualitatively different. Upon increasing pressure, the band near 420 nm attributed to the dication decreases, while a new band around 450 nm emerges, matching the absorption of the radical cation (Fig. 4.52a). A similar behavior is observed for the second aryl-substituted series. These observations show that one-electron reduction of the dication to the radical cation proceeds under hydrostatic pressure in water-containing solvents. A small amount of hydroxide ion present in the medium is considered to act as the reductant. When a monocationic reference salt is examined in a DCM/water mixture under similar pressure conditions, only the usual hyperchromism and bathochromism are observed and no new band associated with a reduced state appears (Fig. 4.52c). Therefore, the monocation remains cationic even under high pressure in the presence of water. Taken together, these comparisons indicate that both the cyclophane architecture and the presence of water are crucial for the pressure-induced one-electron reduction.

Quantitative evaluation of the reaction volume relies on the pressure dependence of the equilibrium between $\mathbf{2}^{2+}$ and $\mathbf{2}^{\bullet+}$. The ratio of the radical cation to dication is treated as an equilibrium constant, and its value is extracted from the overlapping absorption bands using simultaneous equations for the absorbance at wavelengths where each species dominates. A plot of the natural logarithm of the concentration ratio as a function of pressure in the water-containing solvent yields an excellent straight line over most of the pressure range (Fig. 4.53). The slope gives a large negative apparent reaction volume for the reduction. This result supports the initial hypothesis that the volume change consists of a decrease in solvation due to the lower charge and a decrease in the van der Waals volume due to closer π-π stacking when electrostatic repulsion between the xanthylium units is reduced.

DFT provides microscopic support for this interpretation. Geometry optimizations at the (U)ωB97X-D/6-31G* level show that the shortest carbon–carbon distance between the facing π units in the radical cation is shorter by 0.07–0.08 Å compared with the dication (Fig. 4.54). This contraction indicates that the radical cation occupies a smaller van der Waals volume and that the electrostatic repulsion between the cationic π units is indeed reduced upon electron addition. Calculations with an implicit solvent model using the self-consistent reaction field method in DCM and water reveal that the bulk dielectric environment has only a minor effect

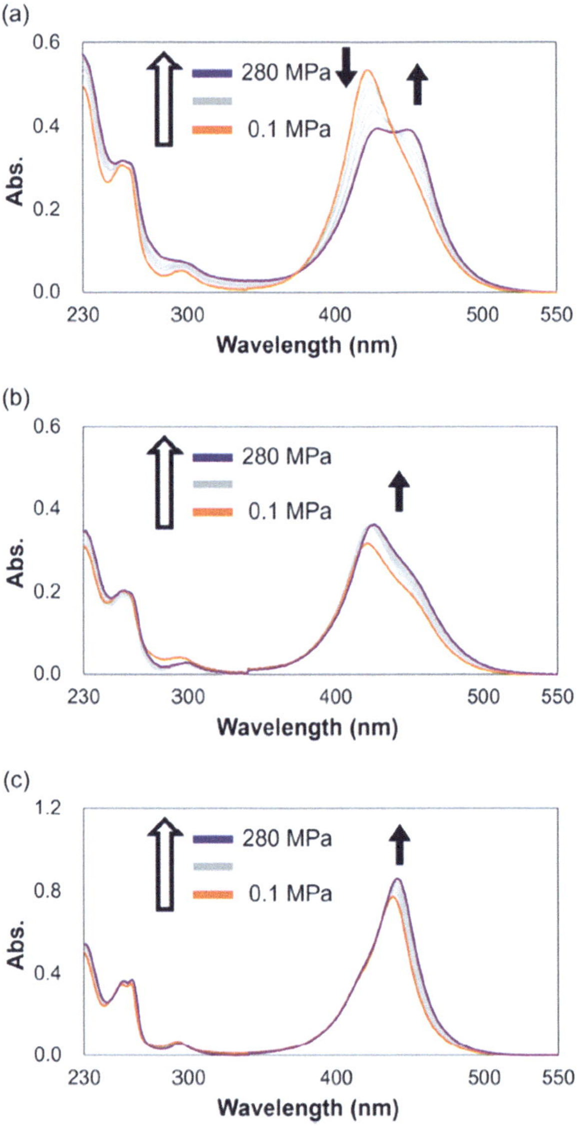

Fig. 4.52 Room-temperature pressure-dependent UV-vis spectra of (**a**) $\mathbf{2a}^{2+}$ $(BF_4^-)_2$ in DCM/H_2O (v/v = 2000/3; $c = 2.90 \times 10^{-5}$ M), (**b**) $\mathbf{2a}^{2+}$ $(BF_4^-)_2$ in anhydrous DCM ($c = 2.82 \times 10^{-5}$ M), and (**c**) $\mathbf{1a}^{+}BF_4^-$ in DCM/H_2O (v/v = 100/1; $c = 4.65 \times 10^{-5}$ M), recorded at 0.1, 40, 80, 120, 160, 200, 240, and 280 MPa using a high-pressure cell (path length 2 mm, medium response, scan rate 100 nm min^{-1}). (Reproduced with permission. Copyright 2025 Royal Society of Chemistry)

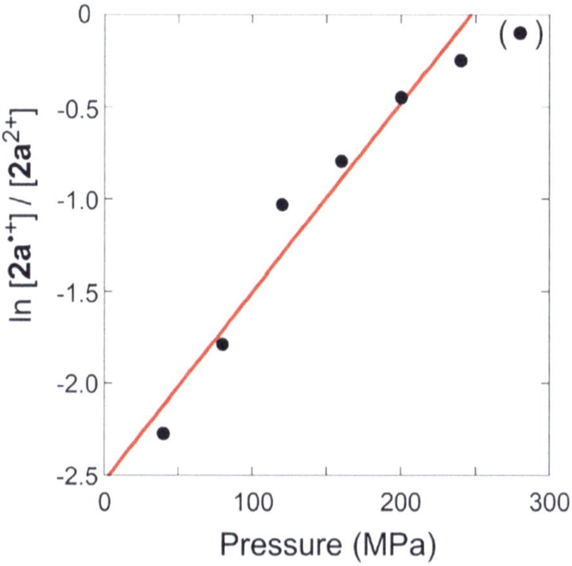

Fig. 4.53 Pressure dependence of the concentration ratio (**2a**$^{2+}$/**2a**$^{\bullet+}$) in DCM/H_2O (v/v = 2000/3) at room temperature (correlation coefficient $r = 0.977$). (Reproduced with permission. Copyright 2025 Royal Society of Chemistry)

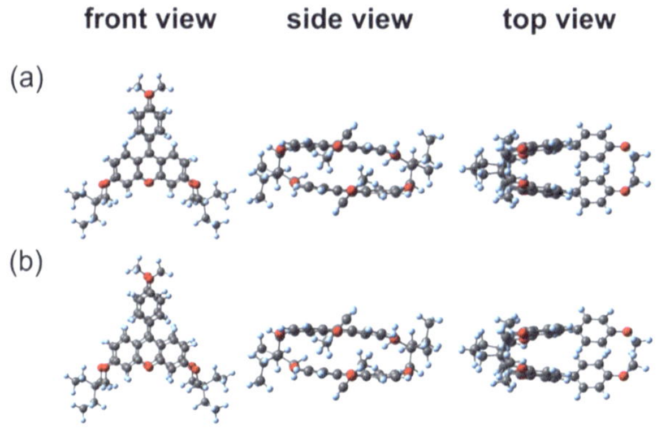

Fig. 4.54 Optimized structures of (**a**) **2a**$^{2+}$ and (**b**) **2a**$^{\bullet+}$ obtained from DFT calculations at the (U) ωB97X-D/6-31G* level under 280 MPa. (Reproduced with permission. Copyright 2025 Royal Society of Chemistry)

on this π-π distance, for both **2**$^{2+}$ and **2**$^{\bullet+}$. The main structural difference between the two redox states is therefore associated with the number of positive charges rather than with direct solvent interactions. The negative reaction volume is consequently attributed to a combination of desolvation of water molecules and reduction of electrostatic repulsion that allows closer intramolecular approach.

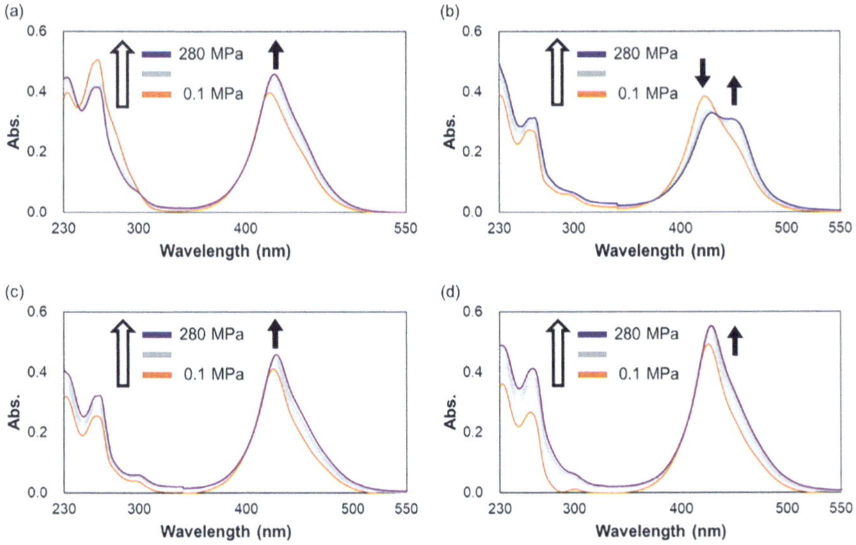

Fig. 4.55 Room-temperature pressure-dependent UV-vis spectra of (**a**) **2a**$^{2+}$(PF_6^-)$_2$ in anhydrous DCM ($c = 2.36 \times 10^{-5}$ M), (**b**) **2a**$^{2+}$(PF_6^-)$_2$ in DCM/H_2O (v/v = 2000:3; $c = 2.29 \times 10^{-5}$ M; containing 3630 equivalents of H_2O per **2a**$^{2+}$), (**c**) **2a**$^{2+}$(NTf_2^-)$_2$ in anhydrous DCM ($c = 1.82 \times 10^{-5}$ M), and (**d**) **2a**$^{2+}$(NTf_2^-)$_2$ in DCM/H_2O (v/v = 2000:3; $c = 2.18 \times 10^{-5}$ M; containing 3820 equivalents of H_2O per **2a**$^{2+}$), recorded at 0.1, 40, 80, 120, 160, 200, 240, and 280 MPa using a high-pressure cell (path length 2 mm, medium response, scan rate 100 nm min^{-1}). (Reproduced with permission. Copyright 2025 Royal Society of Chemistry)

Counter anion effects are probed by measuring pressure dependent spectra for the PF_6^- and NTf_2^- salts in dry and water-containing DCM (Fig. 4.55). In dehydrated solvent, both salts show the same kind of hyperchromic and bathochromic response observed for the BF_4^- salt (Fig. 4.55a, c), consistent with simple pressurization of the solution and preservation of the dicationic state. In the water-containing mixture, the PF_6^- salt behaves analogously to the BF_4^- salt: the dication band decreases and a new band, characteristic of the radical cation, appears with increasing pressure, indicating the occurrence of one-electron reduction (Fig. 4.55b). The NTf_2^- salt, however, continues to display only typical π-conjugated spectral changes in the water-containing solvent and does not show growth of a radical cation band, even at a high pressure (Fig. 4.55d). These results demonstrate that the pressure-responsive redox behavior of the cyclophane dication is strongly influenced by the counter anion.

A solvation-based picture rationalizes this anion dependence (Fig. 4.56). In water-containing DCM, BF_4^- and PF_6^- are less strongly solvated by water than the dicationic cyclophane (Fig. 4.56a). Water molecules therefore preferentially solvate the dication, and hydrostatic pressure induces desolvation of these water molecules upon reduction, which combines with intramolecular compaction to produce a large negative reaction volume and facilitate one-electron reduction. For the more polar NTf_2^- anion, water molecules are expected to solvate the anion more strongly than

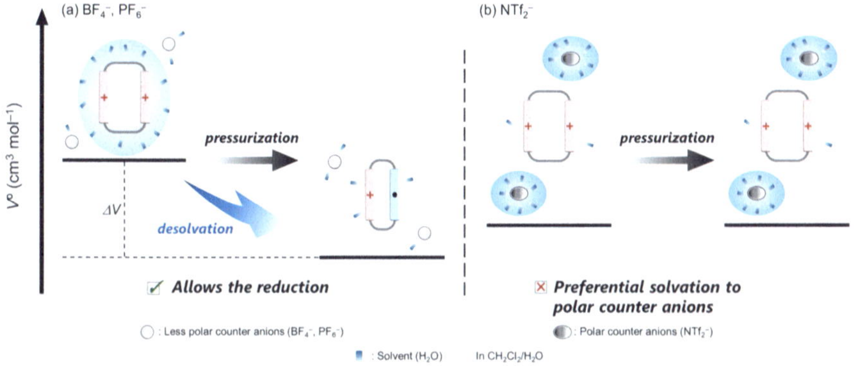

Fig. 4.56 Schematic of (**a**) the volumetric changes accompanying the reduction of **2**$^{2+}$(PF_6^-)$_2$ and **2**$^{2+}$(PF_6^-)$_2$ in DCM/H_2O (v/v = 2000:3), and (**b**) the preferential solvation of the counter anions in **2**$^{2+}$(NTf_2^-)$_2$ under the same solvent conditions. (Reproduced with permission. Copyright 2025 Royal Society of Chemistry)

the dication. In this case, the cyclophane dication is less heavily hydrated, and desolvation does not make a significant contribution when pressure is applied (Fig. 4.56b). The reaction volume becomes much less negative and pressure no longer drives the reduction. This comparison shows that high hydrostatic pressure alone is not sufficient to induce one-electron reduction; the pattern of water solvation around the cation and the anion must cooperate with the cyclophane structure.

In summary, the combined thermodynamic analysis, pressure-dependent spectroscopy, and electronic structure calculations establish that a dicationic cyclophane in a suitable solvation environment can undergo one-electron reduction driven by hydrostatic pressure through a large negative change in apparent reaction volume. Desolvation of water and reduced electrostatic repulsion that allows closer π-π contacts are identified as the key microscopic contributions. Monocationic analogues do not show this behavior, and anion exchange can turn the pressure-responsive redox process on or off by redirecting water solvation. Such results provide concrete design guidelines for molecular systems whose redox behavior can be modulated by hydrostatic pressure and indicate that cationic π-conjugated compounds can be engineered as pressure- and redox-responsive materials by combining cyclophane architectures with controlled solvation environments.

4.12 Pressure-Induced Chiroptical Switching in a Flexible Porphyrin Macrocycle

In this section, hydrostatic pressure functions as an isotropic mechanical input that can reshape molecular conformations and solvation shells via changes in volume (Hamachi et al., 2025). Chiroptical readouts, such as CD and the associated anisotropy factor g, are especially sensitive to subtle helical deformations; therefore, a pressure-driven change in the three-dimensional structure can be directly translated into an optical signal. The system under discussion exemplifies this idea by showing that a single chiral porphyrin macrocycle can invert its chiroptical signature under pressure, switching from a negative to a positive *g* factor in a solvent-dependent and reversible manner.

The porphyrin macrocycle at the center of this study, **DtetMc**, is constructed as a tetrakisporphyrin ring linked by a chiral dioxolane spacer. This architecture combines a twisted, helically chiral framework with pronounced conformational flexibility and known solvent responsiveness. The enantiomer used, with (*R*,*R*) configuration at the dioxolane unit, can adopt either a right-handed (*P*) or left-handed (*M*) twisted conformation, and earlier work showed that the relative stability of these conformers depends sensitively on solvent polarity and structure. The same macrocycle therefore provides a platform where hydrostatic pressure and the solvent microenvironment can act cooperatively or competitively on the *P*/*M* equilibrium and thus on the chiroptical output. An achiral tetrakisporphyrin analogue, **XtetMc**, serves as a reference for separating purely photophysical pressure effects from chiral and conformational contributions (Fig. 4.57).

Baseline measurements on the achiral macrocycle, **XtetMc**, elucidate how a porphyrin tetramer responds to pressure in the absence of helical bias. In toluene and THF, the absorption spectra show a Soret band near 420 nm and Q bands at ~500–700 nm and ambient pressure. When pressure is increased up to several hundred MPa, both Soret and Q bands shift steadily to lower energy at rates on the order of $-1.0\ \mathrm{cm^{-1}\ MPa^{-1}}$, exhibiting very similar values in the two solvents. This behavior is attributed to changes in solvent density and polarizability under pressurization, which stabilize more polar excited states. The overall absorbance increases with pressure, consistent with an effective increase in concentration. Spectra

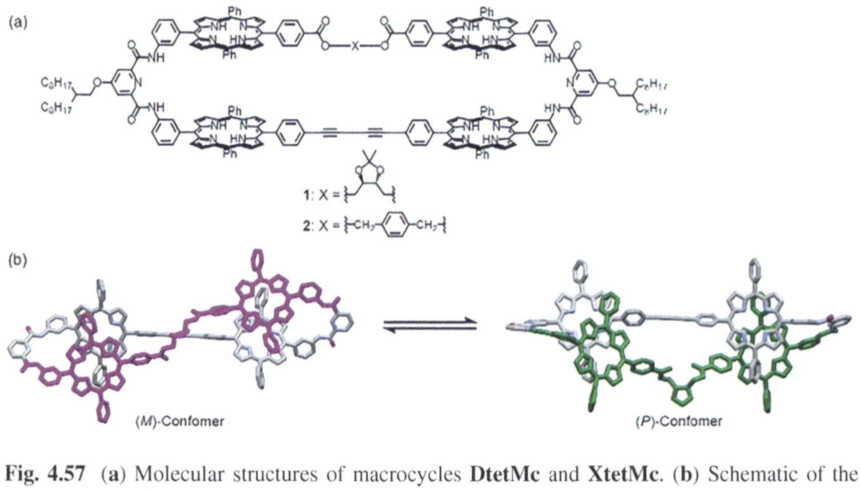

Fig. 4.57 (**a**) Molecular structures of macrocycles **DtetMc** and **XtetMc**. (**b**) Schematic of the pressure-induced chiral inversion of the flexible macrocycle. (Reproduced with permission. Copyright 2025 Royal Society of Chemistry)

recorded after releasing pressure recover the original shapes and positions, confirming reversibility and the absence of irreversible aggregation or structural damage.

Fluorescence measurements of **XtetMc** reveal two emission bands, one at ~600–750 nm assigned to the lowest singlet excited state S_1 of the porphyrin, and another at ~400–500 nm, attributed to emission from a higher excited state S_2. The latter assignment is supported by the excitation-wavelength independence of the S_2 emission maximum. Under pressure, the S_1 emission becomes stronger in both toluene and THF, which is ascribed to the suppression of collisional quenching as solvent viscosity increases. The S_2 band shows solvent-specific trends, decreasing slightly in toluene and increasing more clearly in THF, consistent with different pressure-induced modifications of the S_2 electronic distribution by each solvent. Excitation spectra shift bathochromically with pressure, but fluorescence lifetimes remain essentially unchanged, indicating that the dominant effects are spectral tuning and viscosity-related intensity changes, not fundamental alterations of excited-state decay pathways. Overall, these results confirm that the porphyrin scaffold behaves as a typical π-conjugated chromophore under hydrostatic pressure and that pressure-induced π stacking or macrocycle aggregation does not occur under the conditions used.

The absorption spectra of **DtetMc** in toluene, THF, DCM, and chloroform resemble those of the achiral analogue, with characteristic Soret and Q bands. Emission from both S_1 and S_2 is observed in all solvents, and the excitation spectra show the same type of bathochromic shifts under pressure as in the reference case. Therefore, the underlying chromophore photophysics responds to pressure in a conventional way, independent of chirality. The differences arise in the details of excited-state kinetics and in the chiroptical response.

In non-halogenated solvents, specifically toluene and THF, fluorescence from the S_1 state of the chiral macrocycle increases monotonically with pressure, and the S_1 lifetimes are ~10 ns at ambient pressure and remain essentially constant up to 320 MPa. This pattern mirrors the behavior of the achiral reference. In DCM and chloroform, however, the intensity from S_1 decreases continuously as pressure increases. Lifetimes are already somewhat shorter at ambient pressure, ~8.5 ns, and shorten further to ~7.4 ns at high pressure. These observations suggest that in halogenated solvents the conformational flexibility of the macrocycle enables pressure-induced changes in solvent-mediated quenching pathways in the excited state. The macrocyclic skeleton remains flexible, but the way that flexibility couples to deactivation channels depends strongly on solvent structure and polarity.

The most striking behavior emerges from the CD and *g*-factor profiles. In chloroform and DCM at ambient pressure, the Soret region of **DtetMc** shows a clear positive Cotton effect, which has been assigned previously to the *P* helical conformer (Fig. 4.58a). As pressure increases, the CD band shifts to a lower energy, following the absorption, and the *g* factor near 420 nm grows from a near-zero value to increasingly positive values. This evolution indicates that pressure enhances the chiral response associated with the *P* form without changing its sign. In toluene and THF, the initial CD pattern in the Soret band is a bisignate minus-to-plus couplet, characteristic of the *M* helical conformer (Fig. 4.58b). Application of pressure again causes bathochromic shifts, with the *g* factor around the Soret region evolving from negative values through zero to positive values and continuing to grow in magnitude at high pressures. This sign inversion of the *g* factor under hydrostatic compression is a rare example of a genuine chiroptical switch driven purely by mechanical means.

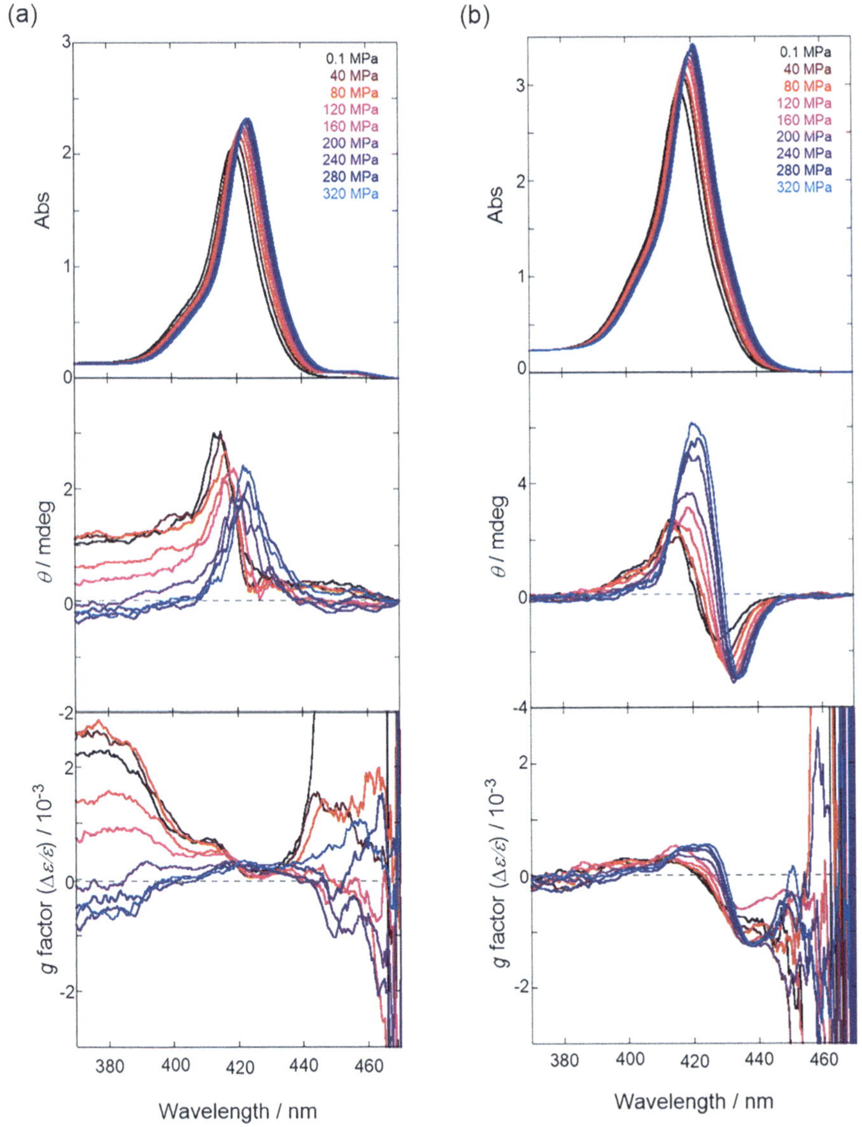

Fig. 4.58 Pressure-dependent UV-vis (top), CD (middle), and anisotropy (*g* factor, $\Delta\varepsilon/\varepsilon$) (bottom) spectra of **DtetMc** in (**a**) chloroform (8 μM) and (**b**) THF (8 μM), recorded at room temperature at 0.1, 40, 80, 120, 160, 200, 240, 280, and 320 MPa (from black to sky blue) using a high-pressure cell. (Reproduced with permission. Copyright 2025 Royal Society of Chemistry)

To quantify this behavior, the pressure susceptibility of the chiroptical response is defined as the slope of g as a function of pressure (Fig. 4.59). Across all four solvents, g increases monotonically with pressure, suggesting that a single overarching mechanism operates (Fig. 4.59a). The underlying picture is that pressurization drives the macrocycle into a more twisted conformation within the same electronic state, exploiting its structural flexibility. The magnitude of the susceptibility, however, depends systematically on the solvent (Fig. 4.59b). In toluene and THF, the slopes are significantly larger than those observed in chloroform and DCM. When susceptibility exceeds a threshold of approximately 1.5×10^{-7} MPa^{-1}, the g factor crosses zero and the system exhibits chiroptical switching behavior. Analysis of the spectral line shapes shows that, in toluene and THF, the g factor at ~420 nm becomes increasingly positive with increasing pressure, while the contribution at longer wavelength decreases. In THF, an additional negative feature near 400 nm also emerges. This spectral evolution is consistent with the characteristic minus-plus-minus Cotton effect sequence associated with the *P* conformer. The pressure-induced evolution of the g spectrum is therefore interpreted as a gradual conversion from the less stable *M* helical form to the more stable *P* form, driven by applied pressure.

The volume origin of this interconversion is attributed to solvation rather than intrinsic molecular size. Structural interconversion between the *P* and *M* forms is not expected to change the intrinsic molecular volume significantly, as both conformers are constructed from the same macrocycle. Instead, the decisive factor is the release or uptake of solvent molecules at the macrocycle surface. In low polarity solvents such as toluene and THF, pressure is expected to expel solvent molecules from around the macrocycle—i.e., to induce desolvation—in order to reduce the total volume of the system. If the *P* form is volumetrically more compact when solvation is taken into account, pressure-induced desolvation will favor its

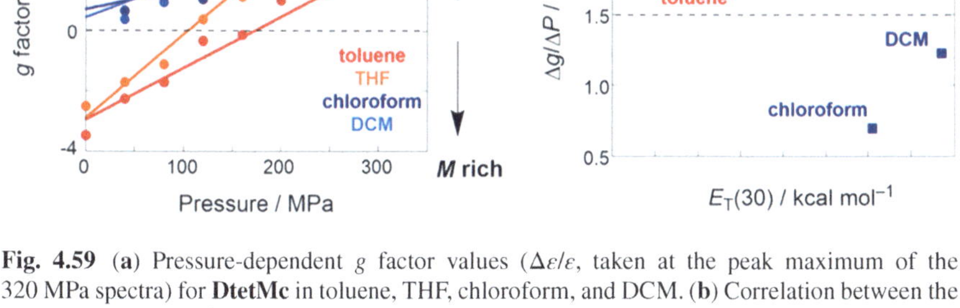

Fig. 4.59 (**a**) Pressure-dependent g factor values ($\Delta\varepsilon/\varepsilon$, taken at the peak maximum of the 320 MPa spectra) for **DtetMc** in toluene, THF, chloroform, and DCM. (**b**) Correlation between the pressure sensitivity of the g factor and $E_T(30)$. The dashed line denotes the threshold for chiroptical switching. (Reproduced with permission. Copyright 2025 Royal Society of Chemistry)

population at the expense of the *M* conformer. Once the *P* conformer becomes dominant, further increasing pressure simply amplifies its chiroptical signal. In chloroform and DCM, where the *P* conformer is already the thermodynamically-preferred species at ambient pressure, the same pressure-induced twisting mechanism operates but only fine-tunes an already *P*-rich ensemble. As a result, the net change in *g* is smaller and no sign inversion occurs.

Taken together, these observations show that a twisted porphyrin macrocycle with a chiral and flexible linker can translate solvent-dependent conformational equilibria into pressure-responsive chiroptical signals. Solvent polarity and structure establish the initial balance between *P* and *M* helices, while hydrostatic pressure shifts this balance through desolvation-driven volume effects. The resulting control over the sign and magnitude of the *g* factor reveals a practical design strategy for pressure-induced chiroptical switches, in which a structurally compliant macrocycle is combined with a solvent environment that allows pressure to reshape the solvation shell and thus the helical conformer distribution.

References

Bovey, F. A., & Yanari, S. S. (1960). Effect of solvent polarizability on the ultra-violet spectral shifts of aromatic compounds. *Nature, 186*, 1042–1044.

Förster, T., & Hoffmann, G. (1971). Die Viskositätsabhängigkeit der Fluoreszenzquantenausbeuten einiger Farbstoffsysteme. *Zeitschrift für Physikalische Chemie, 75*, 63–76.

Hamachi, T., Kinoshita, T., Tanabe, K., Hisano, N., Haino, T., & Fukuhara, G. (2025). A chiroptical switcher: Helical inversion of twisted porphyrin macrocycle synergistically controlled by combined stimuli of solvent and pressure. *Chemical Communications, 61*, 18340–18343.

He, K., Li, Y., Xiang, B., Zhao, P., Hu, Y., Huang, Y., Li, W., Nie, Z., & Yao, S. (2015). A universal platform for building molecular logic circuits based on a reconfigurable three-dimensional DNA nanostructure. *Chemical Science, 6*, 3556–3564.

Hestand, N. J., & Spano, F. C. (2018). Expanded theory of H- and J-molecular aggregates: The effects of vibronic coupling and intermolecular charge transfer. *Chemical Reviews, 118*, 7069–7163.

Hirakawa, K., Saito, K., & Segawa, H. (2009). Anomalously selective quenching of S_2 fluorescence from upper excited state of zinc 5-(1′-pyrenyl)-10,15,20-triphenylporphyrin derivatives through intramolecular charge transfer state. *The Journal of Physical Chemistry. A, 113*, 8852–8856.

Hirakawa, K., Hosokawa, Y., Nishimura, Y., & Okazaki, S. (2019). Relaxation process of S_2 excited zinc porphyrin through interaction with a directly connected phenanthryl group. *Chemical Physics Letters, 732*, 136652.

Jezowski, S. R., Zhu, L., Wang, Y., Rice, A. P., Scott, G. W., Bardeen, C. J., & Chronister, E. L. (2012). Pressure catalyzed bond dissociation in an anthracene cyclophane photodimer. *Journal of the American Chemical Society, 134*, 7459–7466.

Kikuchi, M., Kuwabara, T., Fukuhara, G., Suzuki, T., & Ishigaki, Y. (2025). A redox reaction triggered by hydrostatic pressure in dicationic cyclophanes. *Materials Chemistry Frontiers, 9*, 2863–2870.

Kinoshita, T., Fujise, K., Tsurumaki, E., Toyota, S., & Fukuhara, G. (2022). A pressure-induced ratiometric signalling chemosensor: A case of helical anthracenes. *Chemical Communications, 58*, 3290–3293.

Kinoshita, T., Nakamura, S., Harada, M., Hasobe, T., & Fukuhara, G. (2023). Control of intramolecular singlet fission in a pentacene dimer by hydrostatic pressure. *Chemical Science, 14*, 3293–3301.

Kinoshita, T., Watanabe, K., Tsurumaki, E., Toyota, S., & Fukuhara, G. (2025). Pseudohelicene chemosensor displaying ternary signaling stimulated by hydrostatic pressure and solvent. *Chemical Communications, 61*, 1124–1127.

Kosaka, T., Inoue, Y., & Mori, T. (2016). Toroidal interaction and propeller chirality of hexaarylbenzenes. Dynamic domino inversion revealed by combined experimental and theoretical circular dichroism studies. *The Journal of Physical Chemistry Letters, 7*, 783–788.

Kosaka, T., Iwai, S., Inoue, Y., Moriuchi, T., & Mori, T. (2018). Solvent and temperature effects on dynamics and chiroptical properties of propeller chirality and toroidal interaction of hexaarylbenzenes. *The Journal of Physical Chemistry. A, 122*, 7455–7463.

Kosaka, T., Iwai, S., Fukuhara, G., Imai, Y., & Mori, T. (2019). Hydrostatic pressure on toroidal interaction and propeller chirality of hexaarylbenzenes: Explicit solvent effects on differential volumes in methylcyclohexane and hexane. *Chemistry - A European Journal, 25*, 2011–2018.

Miyagawa, A., Eng, J., Okada, T., Inoue, Y., Penfold, T. J., & Fukuhara, G. (2020). Hydrostatic pressure-induced spectral variation of Reichardt's dye: A polarity/pressure dual indicator. *ACS Omega, 5*, 897–903.

Miyata, K., Conrad-Burton, F. S., Geyer, F. L., & Zhu, X. Y. (2019). Triplet pair states in singlet fission. *Chemical Reviews, 119*, 4261–4292.

Ogawa, R., Kinoshita, T., Kuwabara, T., Sakai, H., Harada, M., Hasobe, T., & Fukuhara, G. (2025). Critical molecular design that can actively control intramolecular singlet fission by hydrostatic pressure. *Chemical Science, 16*, 20245–20254.

Okazaki, M., Takeda, Y., Data, P., Pander, P., Higginbotham, H., Monkman, A. P., & Minakata, S. (2017). Thermally activated delayed fluorescent phenothiazine–dibenzo[*a,j*]phenazine–phenothiazine triads exhibiting tricolor-changing mechanochromic luminescence. *Chemical Science, 8*, 2677–2686.

Ono, S., Kinoshita, T., Iwasaki, H., Imai, Y., & Fukuhara, G. (2024). Ratiometric chemosensors that are capable of quantifying hydrostatic pressure stimulus: A case of porphyrin tweezers. *ACS Physical Chemistry Au, 4*, 510–521.

Ran, X., Pu, F., Ren, J., & Qu, X. (2014). DNA-regulated upconverting nanoparticle signal transducers for multivalued logic operation. *Small, 10*, 1500–1503.

Reichardt, C. (1994). Solvatochromic dyes as solvent polarity indicators. *Chemical Reviews, 94*, 2319–2358.

Reichardt, C., & Welton, T. (2011). *Solvents and solvent effects in organic chemistry*. Wiley.

Sagara, Y., Weder, C., & Tamaoki, N. (2017). Asymmetric cyclophanes permit access to supercooled nematic liquid crystals with stimulus-responsive luminescence. *Chemistry of Materials, 29*, 6145–6152.

Sagara, Y., Tamaoki, N., & Fukuhara, G. (2018). Cyclophane-based fluorescence tuning induced by hydrostatic pressure changes. *ChemPhotoChem, 2*, 959–963.

Smith, M. B., & Michl, J. (2010). Singlet fission. *Chemical Reviews, 110*, 6891–6936.

Takeda, Y., Mizuno, H., Okada, Y., Okazaki, M., Minakata, S., Penfold, T., & Fukuhara, G. (2019). Hydrostatic pressure-controlled ratiometric luminescence responses of a dibenzo[*a,j*]phenazine-cored mechanoluminophore. *ChemPhotoChem, 3*, 1203–1211.

Ullrich, T., Munz, D., & Guldi, D. M. (2021). Unconventional singlet fission materials. *Chemical Society Reviews, 50*, 3485–3518.

Watanabe, K., Tsurumaki, E., Hasegawa, M., & Toyota, S. (2024). Structure and chiroptical properties of anthra[1,2-*a*]anthracene-1-yl dimers as new biaryls. *Chemistry-A European Journal, 30*, e202400929.

Yao, J., Wu, W., Liang, W., Feng, Y., Zhou, D., Chruma, J. J., Fukuhara, G., Mori, T., Inoue, Y., & Yang, C. (2017). Temperature-driven planar chirality switching of a pillar[5]arene-based molecular universal joint. *Angewandte Chemie, International Edition, 56*, 6869–6873.

Yao, J., Mizuno, H., Xiao, C., Wu, W., Inoue, Y., Yang, C., & Fukuhara, G. (2021). Pressure-driven, solvation-directed planar chirality switching of cyclophano-pillar[5]arenes (molecular universal joints). *Chemical Science, 12*, 4361–4366.
Zhou, Z., Wang, J., Levine, R. D., Remacle, F., & Willner, I. (2021). DNA-based constitutional dynamic networks as functional modules for logic gates and computing circuit operations. *Chemical Science, 12*, 5473–5483.

Chapter 5
Pressure-Responsive Supramolecular Chemosensors

Abstract Hydrostatic pressure provides a powerful yet underexploited parameter for regulating supramolecular recognition and signaling in solution. Because pressure directly couples to changes in molar volume, it enables quantitative perturbation of noncovalent interactions, conformational equilibria, solvation structures, and excited-state relaxation pathways that define the function of supramolecular chemosensors. Across diverse host architectures, pressure shifts pre-existing distributions between compact and expanded conformers, alters local polarity and solvation, and modulates the thermodynamics of guest association through characteristic reaction volumes. Chiral assemblies further translate pressure into changes in helicity and stereoselectivity, revealing how solvation and desolvation events govern enantioselective recognition under isotropic compression. In excited states, pressure influences emission color, intensity, and lifetimes by modifying chromophore orientation, packing, and solvent reorganization. Taken together, these observations establish general principles for pressure-responsive supramolecular systems, including volumetric coupling, solvent-driven modulation, conformational adaptability, and reversible reorganization of weak interactions. These concepts provide a unified framework for designing pressure-sensitive supramolecular chemosensors that convert mechanical stimuli into optical or structural outputs.

Keywords Hydrostatic pressure · Supramolecular recognition · Chiral recognition

5.1 Pressure-Modulated Conformational Equilibria and Guest Binding in Dansyl-Modified γ-Cyclodextrin

The design of synthetic receptors whose molecular recognition behavior can be modulated by external stimuli enables responsive sensing and offers mechanistic insight into noncovalent interactions. Hydrostatic pressure is a promising stimulus because it can perturb conformational equilibria and solvent organization in solution without introducing the anisotropies inherent to solid-state methods. A γ-cyclodextrin bearing an *N*-dansyl-D-leucine appendage serves as a convenient

T. Hamachi, G. Fukuhara, *Hydrostatic Pressure Chemistry*,
https://doi.org/10.1007/978-981-95-7461-2_5

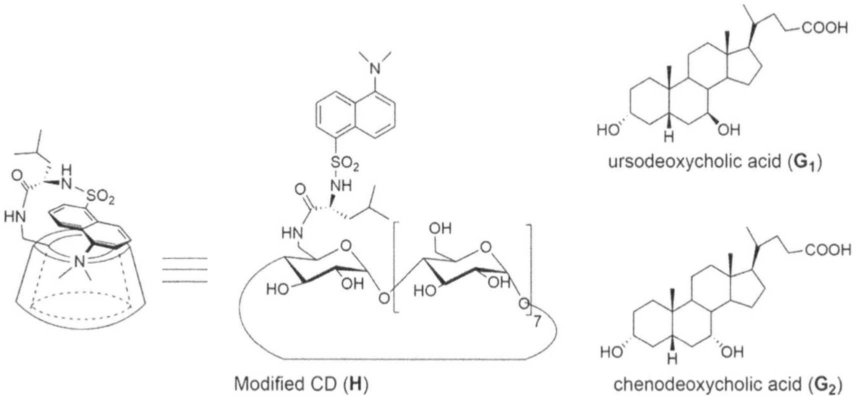

Fig. 5.1 Molecular structures of *N*-dansyl-D-leucine–substituted γ-CD (**H**), ursodeoxycholic acid (**G**$_1$), and chenodeoxycholic acid (**G**$_2$)

reporter that couples a local conformational switch to spectroscopic readout (Fig. 5.1) (Ikeda et al., 1996), because the dansyl chromophore is sensitive to polarity and orientation. The combination of hydrostatic-pressure UV-vis, circular dichroism (CD), steady-state fluorescence, and fluorescence lifetime measurements, together with DFT assignments of dipole directions, reveals how pressure reorganizes the host conformation and, in turn, modulates binding of two diastereomeric bile acids (Fig. 5.1) (Miyagawa et al., 2021). Therefore, such high-pressure spectroscopic analyses provide deeper insight into pressure-dependent chemical sensing using modified cyclodextrins.

CD spectroscopy provides direct evidence for a pressure-dependent pre-equilibrium between two conformers of the modified host, designated naphthyl-in and naphthyl-out, corresponding to the dansyl group positioned inside the cyclodextrin cavity or located on the primary rim, respectively (Fig. 5.2). DFT calculations assign absorptions near 250 and 360 nm to dipole moments along the long and short axes of the naphthalene chromophore (Fig. 5.3). Application of the sector rule indicates that the positive and negative Cotton effects observed at the long and short absorption axes reflect the orientations of the dansyl dipole moments (Fig. 5.3a) (Kodaka, 1993; Harata & Uedaira, 2006). The longer dipole projects along the depth axis of the sector cone, while the shorter dipole resides in the negative sector, consistent with deep penetration of the dansyl substituent into the cyclodextrin cavity (Fig. 5.3b). Upon pressurization up to 20 MPa, the progressively enhanced negative Cotton effects—most notably at 250, 254, and 260 nm—can be ascribed to exclusion of the appended branch from the cavity, because both dipole moments occupy the negative sector region, corresponding to the "naphthyl-out" conformer (Fig. 5.3b). CD spectra recorded from ambient pressure to approximately 280 MPa showed a progressive change in ellipticity in a negative direction up to approximately 20 MPa, followed by saturation above ~40 MPa (Fig. 5.2b). These spectral trends are consistent with gradual exclusion of the dansyl branch from the cavity as

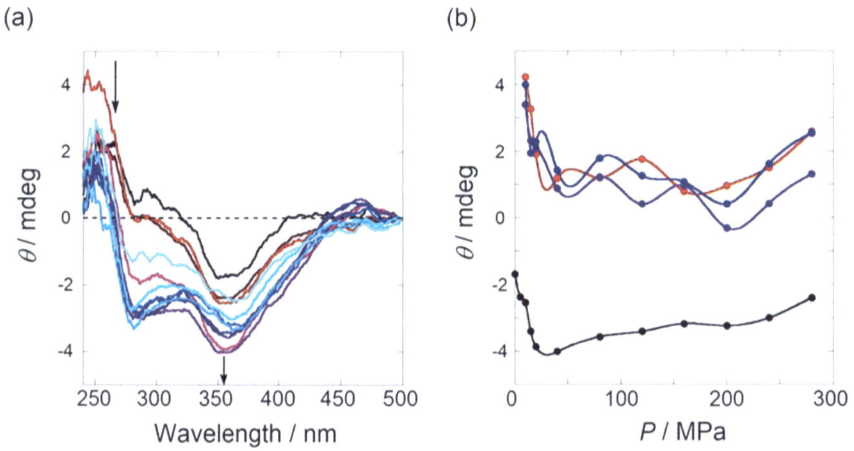

Fig. 5.2 (**a**) CD spectra of **H** (300 μM) recorded in H_2O at room temperature under hydrostatic pressures of 0.1, 5, 10, 15, 20, 40, 80, 120, 160, 200, 240, and 280 MPa (from black to sky blue). (**b**) Pressure-dependent changes in ellipticity of **H** monitored at 250 nm (red), 254 nm (blue), 260 nm (purple), and 360 nm (black). Data at 250 nm for 0.1 and 5 MPa were excluded due to baseline irregularities. (Reproduced with permission. Copyright 2021 John Wiley & Sons)

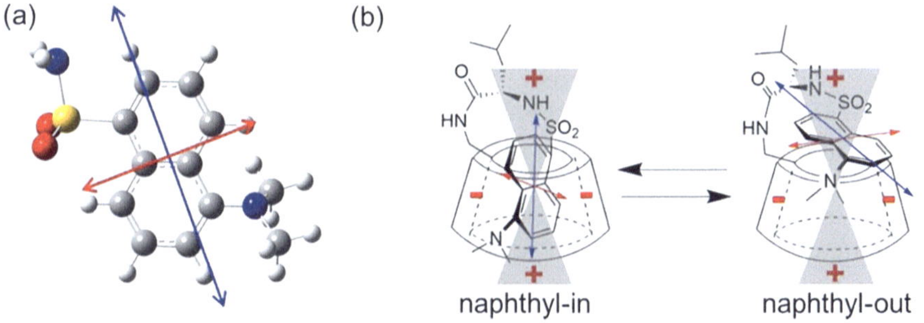

Fig. 5.3 (**a**) DFT-calculated dipole moments of the dansyl chromophore. (**b**) Schematic representations of the "naphthyl-in" and "naphthyl-out" conformations of **H**, illustrated with the sector cone. (Reproduced with permission. Copyright 2021 John Wiley & Sons)

pressure increases, and stabilization of an out conformer at higher pressures. Thus, pressure shifts the equilibrium toward the exposed reporter, altering the population distribution available to engage guest molecules.

Steady-state fluorescence and time-resolved measurements corroborate the CD interpretation and provide additional mechanistic insight. The fluorescence intensity of the host decreases with applied pressure, with a notably steeper decline between 0.1 and 20 MPa followed by a more gradual decrease at pressures above 40 MPa (Fig. 5.4). This behavior contrasts with that of many organic fluorophores, which often increase in brightness under pressure due to higher solvent viscosity, and therefore supports the assignment of emission loss to increased exposure of the

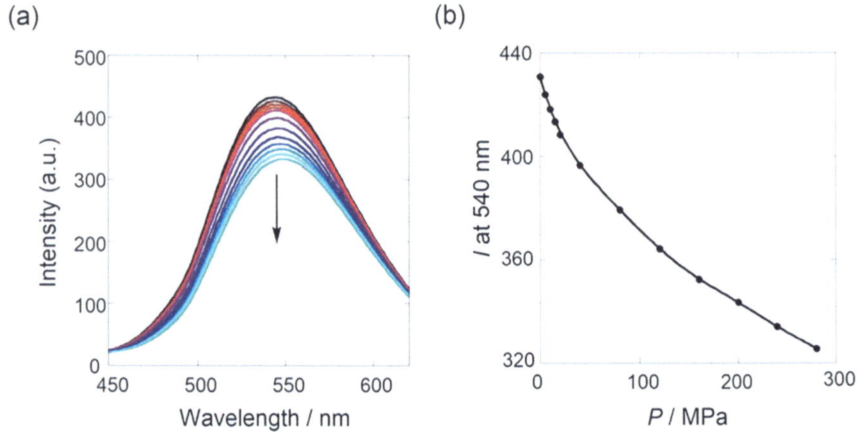

Fig. 5.4 (**a**) Fluorescence spectra of **H** (25 μM) in H_2O at room temperature, excited at 320 nm, under hydrostatic pressures of 0.1, 5, 10, 15, 20, 40, 80, 120, 160, 200, 240, and 280 MPa (from black to sky blue). (**b**) Pressure-dependent variation in fluorescence intensity of **H** monitored at 540 nm. (Reproduced with permission. Copyright 2021 John Wiley & Sons)

dansyl unit to polar bulk water. Emission maxima exhibit modest bathochromic shifts from 543 to 548 nm over the entire pressure range, reflecting both intrinsic pressure effects on the dye and changes in its local environment. Fluorescence decay analysis at 540 nm resolves two components: a longer-lived species near 10.8 ns assigned to the naphthyl-in conformer, where solvent quenching is reduced by encapsulation, and a shorter-lived species near 5–6 ns assigned to the naphthyl-out conformer, where solvent interactions accelerate nonradiative decay. With increasing pressure, the relative amplitude of the shorter component increases and its lifetime decreases slightly, consistent with both a population shift toward the out state and subtle changes in solvent polarity at the reporter site. Notably, lifetime measurements remain sensitive to microenvironmental changes even in the pressure region where the CD signals have plateaued, demonstrating that fluorescence lifetime measurement can detect fine variations in hydrophobicity around the chromophore.

Next, the binding constants (K) of the cholic acids $\mathbf{G_1}$ and $\mathbf{G_2}$ were evaluated as model steroid guests under increasing pressure in H_2O containing ~2 vol.% MeOH by fluorescence titration. As shown in Fig. 5.5, gradual addition of $\mathbf{G_1}$ at 0.1 MPa caused a decrease in fluorescence intensity, indicating that the dansyl group was displaced from the cavity. Assuming a 1:1 binding stoichiometry, nonlinear least-squares fitting (Fig. 5.5b) yielded $K = 20{,}900 \pm 250\ M^{-1}$ for $\mathbf{G_1}$ and $4870 \pm 70\ M^{-1}$ for $\mathbf{G_2}$. These values are lower than those reported by Ueno et al. (24,600 for $\mathbf{G_1}$ and 16,100 for $\mathbf{G_2}$), likely due to solvent differences, as salt-containing buffers were avoided in this study. The difference between $\mathbf{G_1}$ and $\mathbf{G_2}$ arises from their diastereomeric relationship. As shown in Fig. 5.5b, d, high-pressure titrations proceeded smoothly without precipitation.

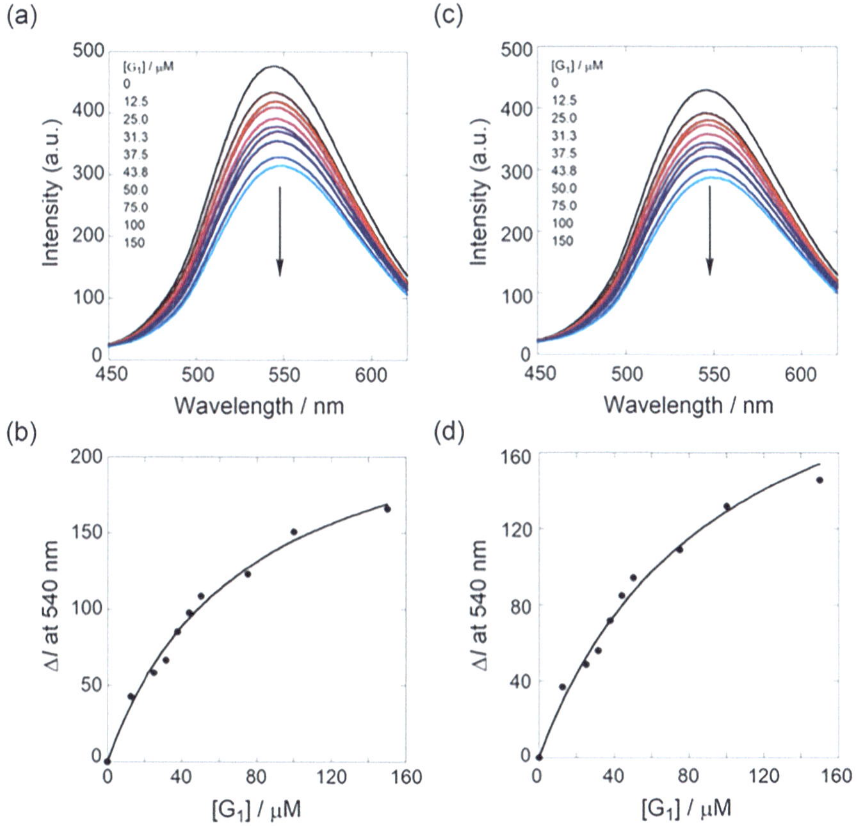

Fig. 5.5 Fluorescence spectral changes of **H** (25 μM) in H_2O containing ~2 vol.% MeOH upon incremental addition of $\mathbf{G_1}$ (0–150 μM) at (**a**) 0.1 MPa and (**c**) 20 MPa at 25 °C, measured using a high-pressure cell. Nonlinear least-squares fitting results assuming a 1:1 binding stoichiometry to determine the binding constants at (**b**) 0.1 MPa and (**d**) 20 MPa. (Reproduced with permission. Copyright 2021 John Wiley & Sons)

The pressure-dependent conformational distribution of the host exerts a direct influence on guest binding. Fluorescence titrations performed in water containing up to 2 vol.% methanol—conditions chosen to avoid salt-induced precipitation under pressure—yield apparent binding constants that differ substantially for the two bile acids. At ambient pressure, the binding constants are approximately 2.09×10^4 M^{-1} for $\mathbf{G_1}$ and 4.87×10^3 M^{-1} for $\mathbf{G_2}$, values that are lower than some literature reports obtained in buffered media (Ikeda et al., 1996), underscoring the solvent sensitivity of these equilibria. Upon pressurization to 20 MPa both binding constants decline modestly but systematically. Linear plots of ln K versus pressure over the low-to-moderate pressure range yield positive reaction volumes (ΔV) of 25.6 cm^3 mol^{-1} for $\mathbf{G_1}$ and 16.1 cm^3 mol^{-1} for $\mathbf{G_2}$ (Fig. 5.6). These positive ΔV values indicate that complexation leads to an overall increase in system volume.

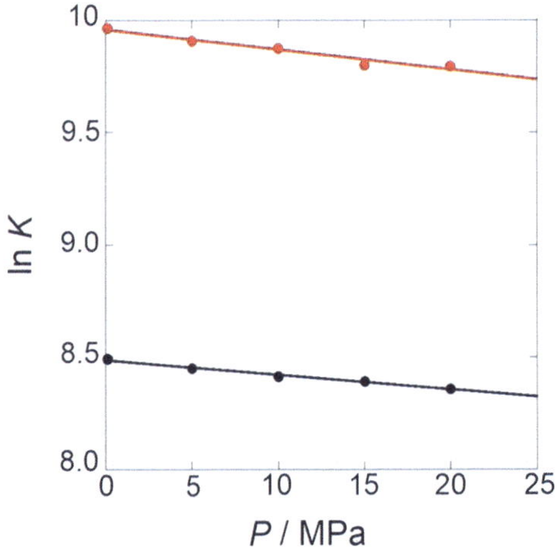

Fig. 5.6 Pressure dependence of ln K for $\mathbf{G_1}$ (red) and $\mathbf{G_2}$ (black), measured at 0.1, 5, 10, 15, and 20 MPa ($r = 0.978$ for $\mathbf{G_1}$ and $r = 0.993$ for $\mathbf{G_2}$). (Reproduced with permission. Copyright 2021 John Wiley & Sons)

Mechanistically, this counterintuitive volumetric expansion is rationalized by the requirement to displace the bulky dansyl appendage from the cavity when a guest enters, together with rearrangement of the solvent molecules around the newly formed complex. Because the pre-equilibrium between in and out conformers shifts with pressure, measurements above the pressure range in which the in conformer is appreciable become unreliable for extracting meaningful pressure dependences of K. Consequently, quantitative analysis is restricted to the early pressurization stage, up to approximately 20 MPa. Within this range, the diastereoselectivity, expressed as the K_{G1}/K_{G2} ratio, remains close to four, reflecting persistent discrimination between the two steroid guests despite the overall decrease in binding affinity.

In summary, hydrostatic pressure induces a redistribution between naphthyl-in and naphthyl-out conformers of *N*-dansyl-D-leucine modified γ-cyclodextrin, and this conformational pre-equilibrium governs both the fluorescence response of fluorophore and the affinity for diastereomeric bile acids. Quantitative analysis in the low-to-moderate pressure range yields positive reaction volumes of 25.6 and 16.1 $cm^3\ mol^{-1}$ for ursodeoxycholic and chenodeoxycholic acids, respectively, indicating that guest binding requires displacement of the bulky dansyl appendage and solvent reorganization rather than simple volume contraction. These results establish that an appended reporter can both signal and mediate pressure-dependent molecular recognition, providing general design principles for pressure-responsive supramolecular sensors in which reporter position and guest accommodation are coupled.

5.2 Protein-Like Pressure-Responsive Guest Binding and Release in a Chiral Supramolecular Polymer

Hydrostatic pressure represents a fundamental thermodynamic parameter that exerts profound effects on the behavior of molecular systems, ranging from small organic molecules to complex biological macromolecules. The study of pressure effects on proteins has provided valuable insights into their adaptability to extreme environments, such as the deep sea, the role of pressure in prebiotic chemistry, and the modulation of enzymatic or binding functions (Heremans, 1982). Proteins typically engage in substrate recognition and binding through non-covalent interactions at well-defined sites, relying on hydrogen bonding, ionic interactions, π-π stacking, and hydrophobic interactions. These interactions are inherently sensitive to structural perturbations, solvation changes, and conformational adjustments, which can be induced by hydrostatic pressure. Consequently, protein binding affinities vary under elevated pressures, manifesting either enhanced association or release of bound substrates, depending on the specific protein-ligand system. Beyond biological macromolecules, artificial molecular systems have also been investigated for their pressure-dependent non-covalent interactions, offering controlled platforms to examine molecular recognition under tunable external conditions. Such studies have encompassed receptor molecules at interfaces, where conformational adaptation under pressure allows selective guest binding, as well as supramolecular polymers exhibiting pre-organized recognition sites that mimic protein binding pockets. These artificial systems serve as simplified models for probing fundamental principles underlying molecular recognition, cooperative interactions, and environmental responsiveness in confined architectures.

In this context, this section focuses on a supramolecular polymer constructed from a bichromophoric naphthalenediimide (NDI) derivative capable of chiral recognition and pressure-responsive behavior. The NDI derivative, available as enantiopure (*R*)-**4** or (*S*)-**6**, contains core-substituted units that self-assemble into nanofibers, forming semi-open pocket-like binding sites formed by bulky ethoxy substituents (Fig. 5.7) (Sethy et al., 2017, 2018). These pockets exhibit a well-defined chiral architecture arising from a one-handed twist induced by a cyclohexyl-1,2-diamide framework, enabling highly selective recognition of

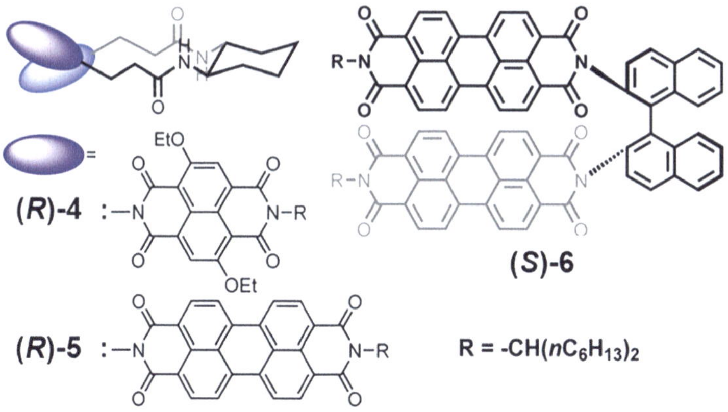

Fig. 5.7 Chemical structures of the enantiomeric (*R*) nanofiber-forming compound (*R*)-**4** and the chiral guest molecules (*R*)-**5** and (*S*)-**6**. (Reproduced with permission. Copyright 2019 Royal Society of Chemistry)

complementary perylenediimide (PDI) guest molecules such as (*S*)-**6** or (*R*)-**5**, respectively (Fig. 5.7). Binding events are sensitively detected through the energy transfer from the NDI units in the nanofibers to the PDI guest, producing a characteristic emission that reflects the efficiency of guest incorporation. The semi-open binding pockets, stabilized by π-π interactions and sterically guided by the chiral scaffold, resemble the clefts found in protein structures, suggesting that they could exhibit comparable sensitivity to hydrostatic pressure (Yonezawa et al., 2019).

Initial investigations evaluated the absorption spectra of the (*R*)-**4**/(*S*)-**6** mixture under pressurization in a chloroform/MCH (1:9) solvent mixture. Concentrations were selected to ensure robust self-assembly, as indicated by the broadening of the π-π* transition around 335 nm. UV-vis measurements under hydrostatic pressure revealed that the overall self-assembly of (*R*)-**4** was preserved across the pressure range, with the π-π* band maintaining its spectral signature, indicating the structural stability of the nanofibers (Fig. 5.8a). However, continuous redshifts were observed in both the π-π* and CT bands, with the latter exhibiting a more pronounced shift (Fig. 5.8b). These spectral changes can be attributed to a combination of factors, including pressure-induced solvent density effects, altered dielectric polarizability, and modulation of inter-chromophore interactions. The observed redshift in the CT band particularly indicates enhanced interaction between quadrupole moments in the donor-substituted NDI units, suggesting subtle but meaningful changes in molecular packing within the nanofiber assembly.

Fluorescence studies further elucidated the pressure-dependent guest binding behavior. Excitation at 410 nm, where absorbance changes were minimal, revealed efficient energy transfer from the NDI units to the bound PDI guest, as evidenced by a strong PDI emission at 540 nm and suppressed NDI emission near 480 nm (Fig. 5.8c). Notably, the ratio of PDI to NDI emission intensities (I_{PDI}/I_{NDI}) initially increased upon pressurization up to 50 MPa, indicating a slight enhancement in energy transfer efficiency (Fig. 5.8d). Beyond this point, however, the emission ratio decreased continuously, suggesting the progressive suppression of energy transfer (Fig. 5.8d). This effect was attributed either to the release of guest molecules from the host pockets or to perturbations in the molecular packing necessary for efficient energy migration. Control experiments using a structurally distinct guest, (*S*)-**5**, which associates with the supramolecular polymer via parallel hydrogen bonding, exhibited pressure-independent energy transfer, supporting the conclusion that the observed changes for (*S*)-**6** reflect actual guest release rather than general packing disruption.

To gain insight into the pressure-dependent binding and release of guest (*S*)-**6** on the (*R*)-**4** assembly, fluorescence anisotropy (r) measurements were performed under pressure. At ambient pressure, the (*R*)-**4**/(*S*)-**6** complex showed a higher r value (0.011) than the mismatched (*R*)-**4**/(*R*)-**6** pair (4.1×10^{-3}) and monomeric (*S*)-**6** (2.9×10^{-3}), indicating guest incorporation (Fig. 5.9). Upon pressurization, r increased gradually for the mismatched pair, reflecting enhanced self-assembly of (*R*)-**6**. For the (*R*)-**4**/(*S*)-**6** system, r stayed constant up to 50 MPa, then decreased to 4.1×10^{-3} at 150 MPa, matching the monomeric (*S*)-**6** value, and approached that of the mismatched pair above 200 MPa. These results suggest that (*S*)-**6** is released

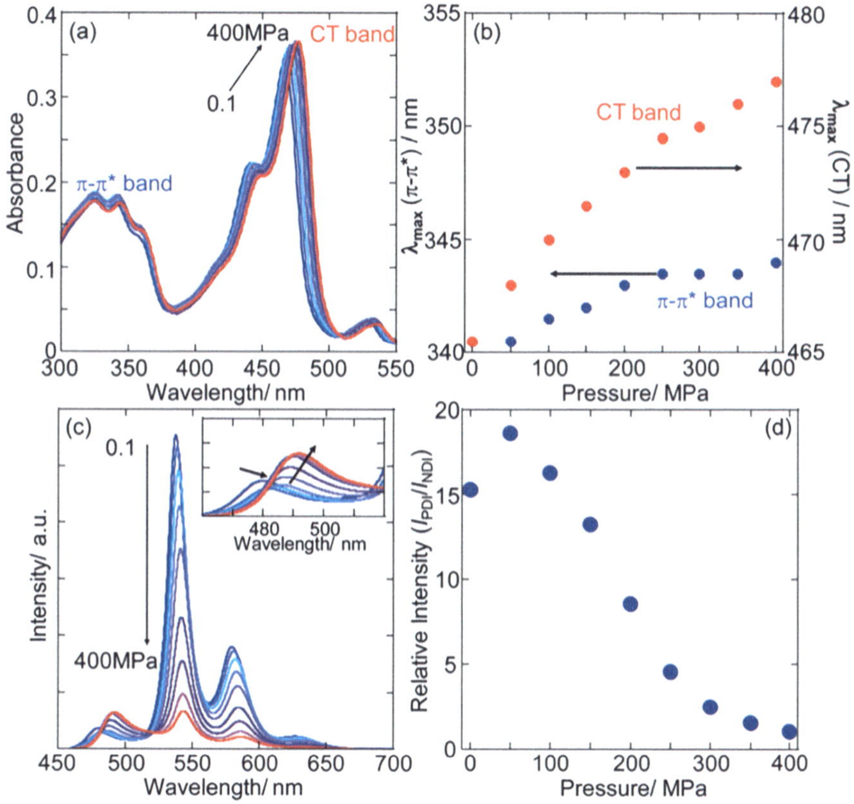

Fig. 5.8 (**a**) Absorption and (**c**) fluorescence spectral variations of (*R*)-**4** in an MCH/chloroform (9:1) mixture containing 2.5 mol% of (*S*)-**6** under hydrostatic pressure. [**4**] = 3.0 × 10^{-5} M, [**6**] = 7.5 × 10^{-7} M, λ_{ex} = 410 nm. Inset in (**c**): Enlarged spectra in the donor emission region (460–520 nm). (**b**) Plots of the absorbance maxima (λ_{max}) of the π-π* and CT bands obtained from (**a**). (**d**) Pressure dependence of the relative emission intensity ratio (I_{PDI}/I_{NDI}). (Reproduced with permission. Copyright 2019 Royal Society of Chemistry)

from the (*R*)-**4** host above 50 MPa and subsequently undergoes homo-assembly at higher pressures.

The binding affinity between the host and guest, quantified through apparent association constants derived from Stern-Volmer analysis of fluorescence quenching, was highly influenced by pressure (Fig. 5.10a). At ambient conditions, the association constant for (*S*)-**6** binding to (*R*)-**4** was 4.9 × 10^6 M^{-1}, exceeding the self-association constant of the guest (Fig. 5.10b). The binding constant exhibited a slight increase up to 50 MPa, followed by a continuous decrease to 1.6 × 10^5 M^{-1} with further pressurization. Such a behavior reflects a non-monotonic volume response, characterized by an apparent negative volume change at low pressures that facilitates binding and a positive volume change at higher pressures that promotes guest release. Similar phenomena have been reported for protein-substrate

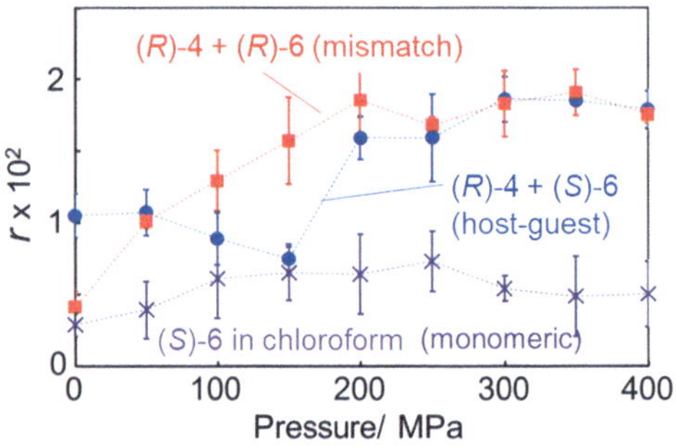

Fig. 5.9 Pressure dependence of fluorescence anisotropy (r). Concentrations: [(R)-**4**] = 3.0×10^{-5} M; [**6**] = 7.5×10^{-7} M. (Reproduced with permission. Copyright 2019 Royal Society of Chemistry)

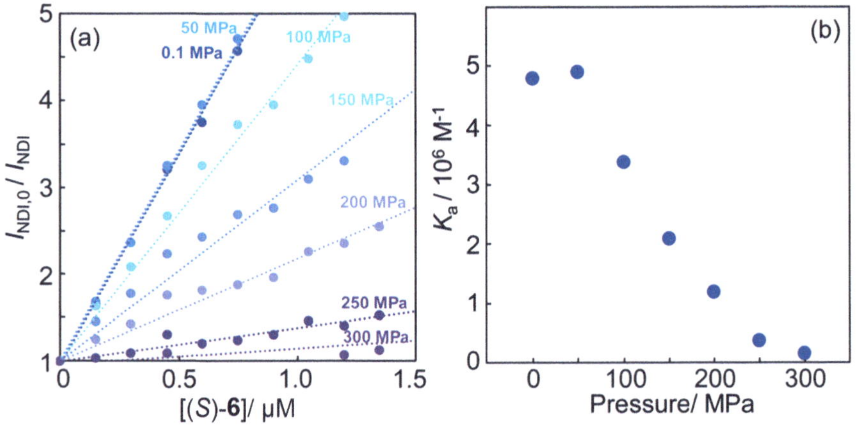

Fig. 5.10 (**a**) Stern-Volmer plots of the (R)-**4** and (S)-**6** host-guest system at different pressures. Stern-Volmer equation: $I_{NDI,0}/I_{NDI} = 1 + K_S[(S)\text{-}\mathbf{6}]$, where $I_{NDI,0}$ is the fluorescence intensity in the absence of (S)-**6** and K_S is the Stern-Volmer constant. (**b**) Pressure dependence on the binding constant K_a. (Reproduced with permission. Copyright 2019 Royal Society of Chemistry)

systems, such as the riboflavin binding protein-flavin mononucleotide complex, where low-pressure stabilization is followed by pressure-induced denaturation and ligand release. In the present supramolecular polymer system, the initial pressure range enhances π-π interactions within the binding pockets, promoting guest association. At higher pressures, however, hydrogen-bond-driven polymerization of the host and self-association of the guest are reinforced, leading to pressurization of binding pockets and eventual release of the guest molecules. The interplay between pressure-modulated host-guest interactions and independent self-association of the host and guest molecules underlies the observed protein-like binding-release response.

In conclusion, a supramolecular polymer constructed from a chiral bichromophoric NDI derivative demonstrates protein-like, pressure-dependent guest binding

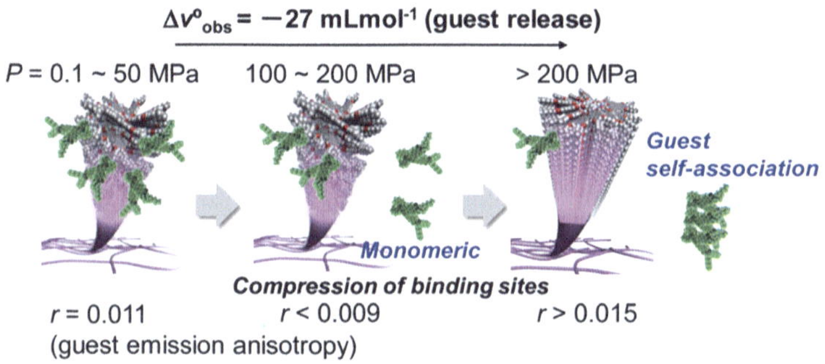

Fig. 5.11 Schematic of the pressure-induced guest binding and release between the (*R*)-**4** host assembly and the (*S*)-**6** guest. (Reproduced with permission. Copyright 2019 Royal Society of Chemistry)

and release. The semi-open binding pockets, stabilized by π-π interactions and chiral arrangement, facilitate selective guest recognition under ambient conditions. Upon pressurization, guest association is initially promoted; however, further increases in pressure reinforce host self-assembly and guest self-association, compressing the binding pockets and ultimately leading to the sequential release of the guest (Fig. 5.11). Fluorescence and absorption spectroscopy, fluorescence anisotropy measurements, and binding constant analyses collectively reveal a non-monotonic response to pressure, characterized by an initial binding facilitation followed by release at higher pressures. These findings highlight the capacity of hydrostatic pressure to modulate non-covalent interactions in a controlled and reversible manner, translating isotropic stress into anisotropic molecular effects, and offer insights into the design of artificial systems that mimic the adaptive, responsive behavior of biological macromolecules. The interplay between directional interactions, supramolecular packing, and guest recognition under pressure provides a conceptual framework for the development of responsive molecular materials and contributes to understanding the fundamental principles governing protein-like functions in synthetic assemblies. This study underscores the potential of pressure as a tunable external stimulus for controlling molecular recognition, energy transfer, and dynamic self-assembly, with implications for both fundamental supramolecular chemistry and the design of functional materials such as molecular sensors, switches, and micromachines.

5.3 Pressure-Modulated Conformational and Chiroptical Behavior of a Fluorescent Foldameric Anion Receptor

Stimulus-responsive supramolecular architectures offer routes to adaptive sensing, controlled delivery, and switchable information storage by converting external inputs into molecular-level structural and functional changes. Hydrostatic pressure, an isotropic mechanical stimulus that directly modulates molar volumes and solvation shells, remains comparatively underexploited for such purposes despite its long-standing use in physical chemistry and biophysics. The present study focuses on the pressure-dependent spectroscopic and binding behavior of a fluorescence foldamer, denoted **H** (Haketa et al., 2012; Haketa & Maeda, 2011), examined as a pressure-manipulable anion receptor in combination with chiral and achiral ion pairs (Fig. 5.12a) (Kinoshita et al., 2021). Chiral guests were provided by binaphthylammonium salts in enantiomerically pure forms, **SS** and **RR**, paired with chloride or bromide anions, whereas tetrabutylammonium salts supplied achiral ion pairs (Fig. 5.12b). The experimental approach combined steady-state UV-vis absorption, fluorescence, and CD spectroscopy with fluorescence lifetime

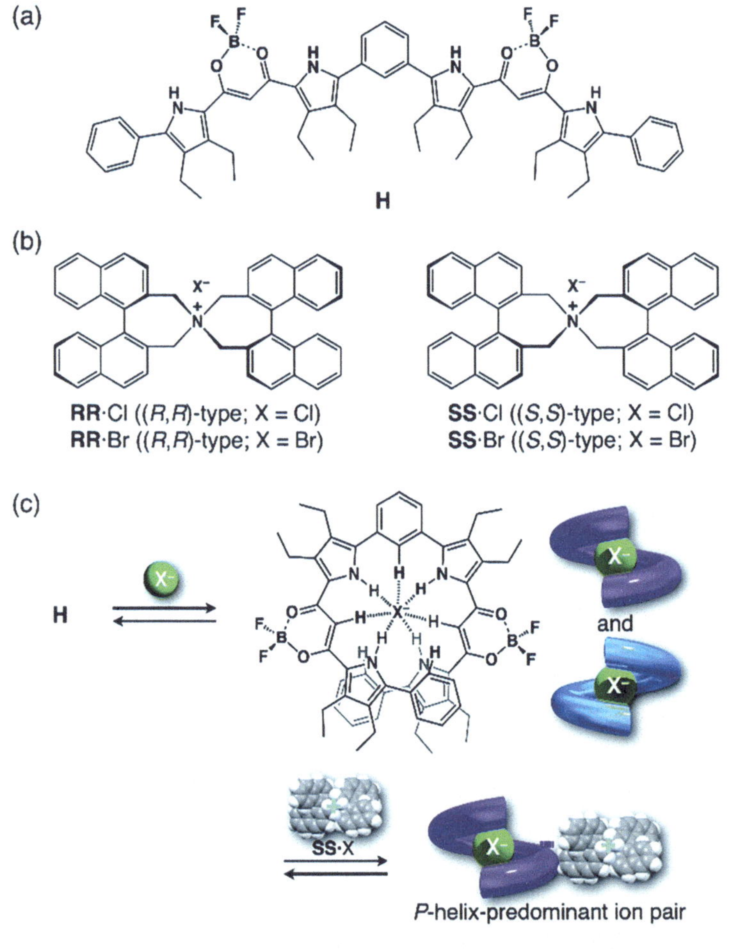

Fig. 5.12 (**a**) Fluorescent anion receptor (**H**, host), (**b**) chiral ion pairs (**SS** or **RR**·X; X = Cl, Br), and (**c**) their supramolecular complexation modes. (Reproduced with permission. Copyright 2021 Royal Society of Chemistry)

measurement and quantitative titrations under hydrostatic pressures ranging from atmospheric conditions to 320 MPa. This strategy enabled simultaneous interrogation of ground-state conformational equilibria, excited-state dynamics, and molecular recognition thermodynamics (Fig. 5.12c).

Spectroscopic monitoring revealed pressure-induced bathochromic shifts and hyperchromic effects in the UV-vis absorption of **H** across several organic solvents, including toluene, chloroform, dichloromethane (DCM), and acetonitrile (Fig. 5.13). Slope parameters for the 0–0 absorption band (α_A) were close to values typical for common organic dyes in nonpolar and moderately polar solvents, approximately $-1.0\ cm^{-1}\ MPa^{-1}$ in toluene, chloroform, and DCM. A larger red shift with pressure was observed in acetonitrile, where α_A reached approximately $-1.34\ cm^{-1}\ MPa^{-1}$ up to ~160 MPa. The absorption and excitation spectra returned to their original forms upon depressurization, confirming the reversibility of the pressure-induced changes. Two optical signatures provided key insight into conformational equilibria. First, changes in the relative intensities of the 0–0 and 0–1 vibronic bands reflected variations in conjugation length and folding. Second, comparison of the pressure-induced spectral shifts between absorption and fluorescence bands, enabled inference of solvent- and pressure-dependent conformational responses.

Accumulated evidence supports the coexistence of an extended conformer (**E**) and a folded conformer (**F**) in the ground state, with solvent polarity governing the baseline distribution. In the relatively polar solvent acetonitrile, spectral features at low pressure resembled those produced upon anion binding, consistent with a predominance of the folded, conjugation-shortened conformer at ambient conditions. Stepwise application of hydrostatic pressure shifted the equilibrium toward the extended state, as manifested by systematic increases in the 0–0/0–1 band ratio and by more pronounced pressure-dependent spectral changes (Fig. 5.14). This pressure-induced shift is rationalized in terms of solvation energetics and volumetric considerations; molecules in the extended conformation present larger solvated surface areas, which are released more readily upon pressurization than those of the more compact folded conformer. Consequently, the conformational populations of the flexible foldamer can be manipulated by changing the hydrostatic pressure.

The fluorescence behavior under pressure displayed two principal features: first, emission spectra showed bathochromic shifts analogous to those in absorption. Second, the emission intensities decreased with increasing pressure, a behavior contrary to expectations based on the Förster-Hoffmann relation, which predict enhanced fluorescence under pressurization due to increased solvent viscosity suppressing nonradiative collisional deactivation. A model consistent with the experimental data proposes an initially relatively planar Franck-Condon state that, under compressed conditions, relaxes into a twisted excited state along a low-energy coordinate. Solvent reorientation accompanying this planar-to-twisted relaxation enables access to a nonradiative decay channel that offsets any viscosity-induced suppression of collisional quenching, thereby yielding net fluorescence quenching with increasing pressure.

Fluorescence lifetime measurements via time-correlated single photon counting supported the conformational interpretation and revealed excited-state

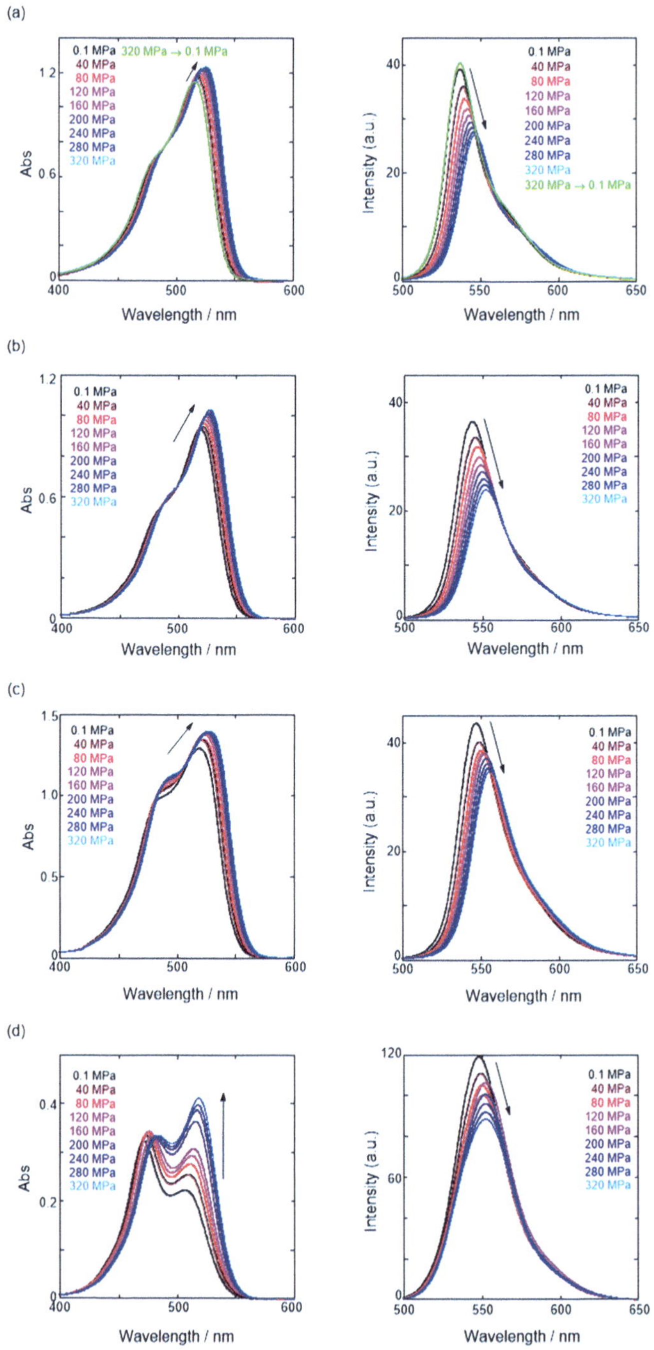
(a)
0.1 MPa
40 MPa
80 MPa
120 MPa
160 MPa
200 MPa
240 MPa
280 MPa
320 MPa
320 MPa → 0.1 MPa
Abs
Wavelength / nm
Intensity (a.u.)
(b)
(c)
(d)

heterogeneity. In toluene, chloroform, and DCM, single exponential decays with lifetimes close to 1.7–1.9 ns were observed at elevated pressures, indicating emission dominated by a single excited species, assigned to the extended conformer, **E***. Conversely, acetonitrile solutions exhibited biexponential decay under pressure, with a short lifetime component of roughly 0.6–0.7 ns and a long lifetime component of approximately 6.5–8.1 ns. The relative amplitudes evolved systematically with pressure, showing an increasing contribution from the shorter-lived species as pressure increased, and hence the short and long components were assigned to **E*** and **F***, respectively. Taken together with steady-state spectroscopic data, these lifetime distributions establish that both ground-state and excited-state conformational populations are responsive to hydrostatic pressure and solvent composition.

Quantitative thermodynamic analysis of anion binding employed UV-vis titrations under controlled pressure to extract the stoichiometric binding constant, K_{anion}, for 1:1 complexation of **H** with the various ion pairs (Fig. 5.15). Nonlinear least-squares fitting of spectral changes provided K_{anion} values at multiple pressures, and linear plots of $\ln K_{anion}$ versus pressure yielded apparent reaction volume changes, ΔV_{app}, via standard thermodynamic relationships (Fig. 5.17). Distinct trends emerged depending on the countercation identity. For binaphthylammonium salts, the ΔV_{app} values were positive and appreciable, with **RR** chloride exhibiting the largest expansion-like volume change upon complexation, approximately +9.1 cm^3 mol^{-1} at 298 K, whereas **SS** bromide showed a smaller positive change near +2.6 cm^3 mol^{-1}. In contrast, tetrabutylammonium salts displayed small negative ΔV_{app} values, approximately −1.5 and −0.3 cm^3 mol^{-1} for chloride and bromide, respectively. The pressure-dependent K_{anion} values decreased with increasing pressure for chiral guests, consistent with shifts in ion pairing and solvation.

The interpretation of these volumetric signatures highlights the role of solvation and the nature of ion pairing. Microscopic contributions to ΔV_{app} were decomposed conceptually into intrinsic van der Waals volumes and solvent-associated molar volumes. Because the intrinsic volume of the foldamer was not expected to change markedly upon complexation, solvent-dependent contributions dominate. A plausible mechanistic picture is that formation of a solvent-shared ion pair that is volumetrically expanded relative to other association modes produces positive ΔV_{app} values, whereas tighter ion pairs or complexes that reduce solvent-exposed volumes lead to negative changes. For binaphthylammonium salts, formation of complexes with expansive solvation shells was supported experimentally by spectroscopic probes.

Fig. 5.13 UV-vis absorption (left) and fluorescence (right, excitation at 404 nm) spectra of **H** recorded at 0.1, 40, 80, 120, 160, 200, 240, 280, and 320 MPa (from black to sky-blue lines) in (**a**) toluene (44 μM), (**b**) chloroform (42 μM), (**c**) DCM (49 μM), and (**d**) acetonitrile (15 μM) at room temperature, measured in a high-pressure cell. The green line represents the spectrum at 0.1 MPa after depressurization from 320 MPa. The absorbance at the excitation wavelength (404 nm) was identical in all measurements. (Reproduced with permission. Copyright 2021 Royal Society of Chemistry)

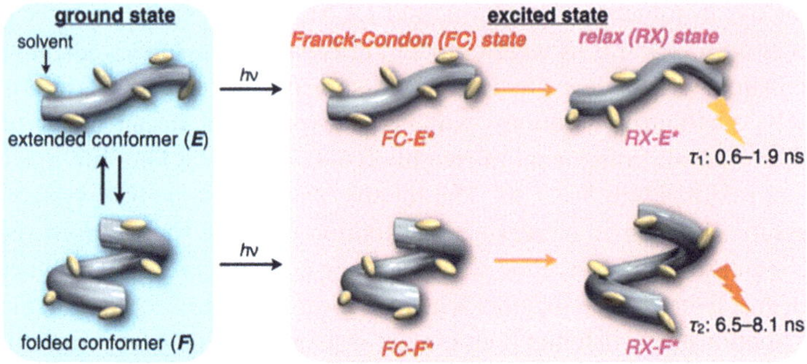

Fig. 5.14 Schematic of the dynamic behavior of **H** in its ground and excited states. (Reproduced with permission. Copyright 2021 Royal Society of Chemistry)

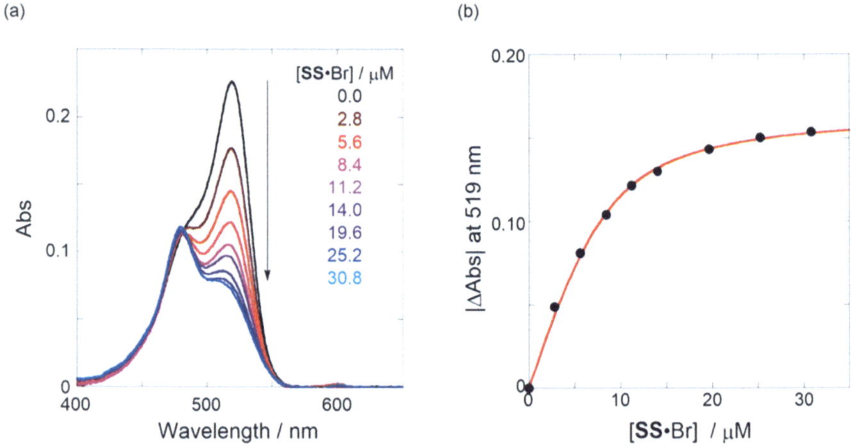

Fig. 5.15 (**a**) UV-vis absorption spectral variations of a chloroform solution of **H** (8.0 μM, black line) upon gradual addition of **SS**·Br (2.8–30.8 μM, colored lines) at room temperature under 40 MPa. (**b**) Nonlinear least-squares fitting of the titration data, assuming a 1:1 binding stoichiometry between **SS**·Br and **H**, to determine the binding constant (K_{anion}) at 40 MPa. (Reproduced with permission. Copyright 2021 Royal Society of Chemistry)

Chiroptical measurements provided complementary insight into the pressure-induced changes in ion pairing and helicity induction. CD spectra obtained for **H** in the presence of enantiopure binaphthylammonium salts showed characteristic anisotropy factors, *g*, that decreased progressively with pressure even when host occupancy remained essentially quantitative (Fig. 5.16). The decline in *g*, accompanied by bathochromic shifts, indicates pressure-promoted conversion of tightly bound, strongly chiral ion pairs toward more weakly associated conformers that induce reduced helicity in the foldamer. Comparison between chloride and bromide counterions showed a steeper pressure-dependent decrease of *g* for chloride, consistent with its stronger initial association and greater sensitivity to pressure-mediated solvation changes.

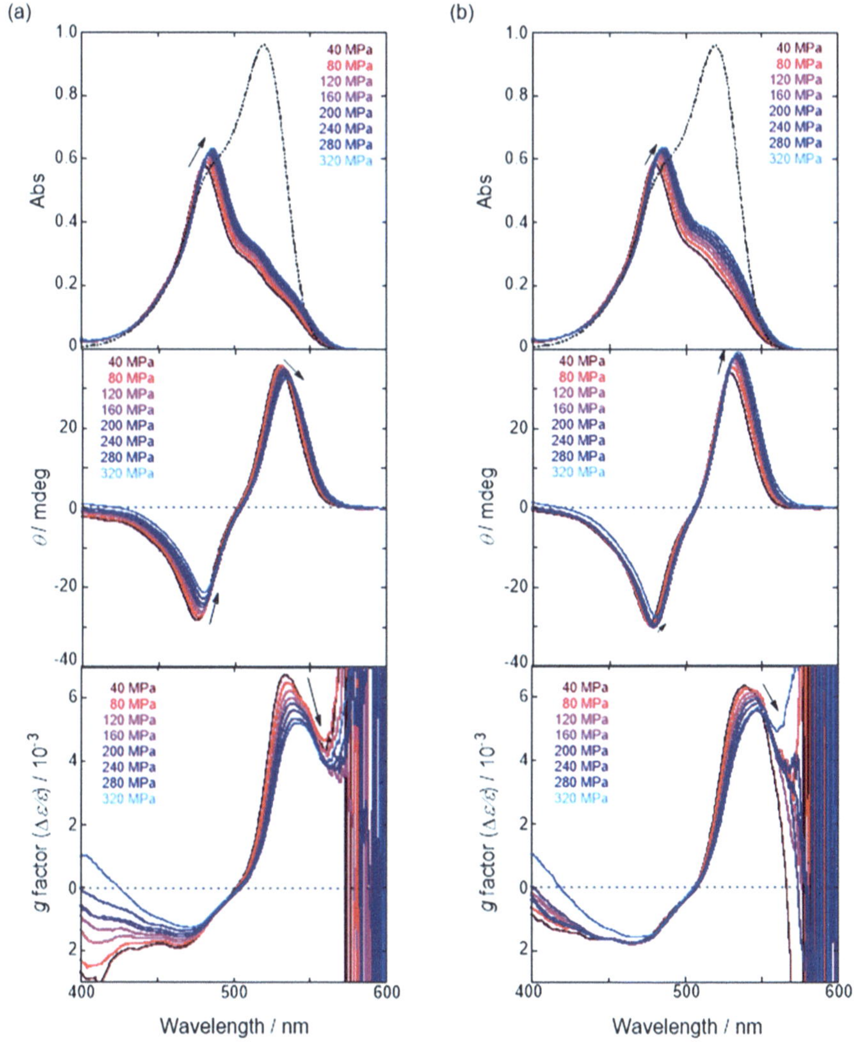

Fig. 5.16 Pressure-dependent UV-vis absorption (top), CD (middle), and anisotropy (g factor = $\Delta\varepsilon/\varepsilon$) (bottom) spectra of **H** (34 μM) in the presence of (**a**) **SS**·Cl (120 μM) and (**b**) **SS**·Br (130 μM) in chloroform at room temperature, recorded at 40, 80, 120, 160, 200, 240, 280, and 320 MPa in a high-pressure cell. The dotted lines represent the UV-vis absorption spectra of **H** (34 μM) without the corresponding guest at 0.1 MPa. (Reproduced with permission. Copyright 2021 Royal Society of Chemistry)

A final set of experiments examined the influence of adding small amounts of a polar cosolvent, methanol, to chloroform solutions. Incremental methanol contents up to about 0.3% did not alter the ΔV_{app} values for both chloride and bromide, whereas K_{anion} decreased with increasing methanol content (Fig. 5.17). This combination of invariance in the apparent reaction volume with decreasing binding

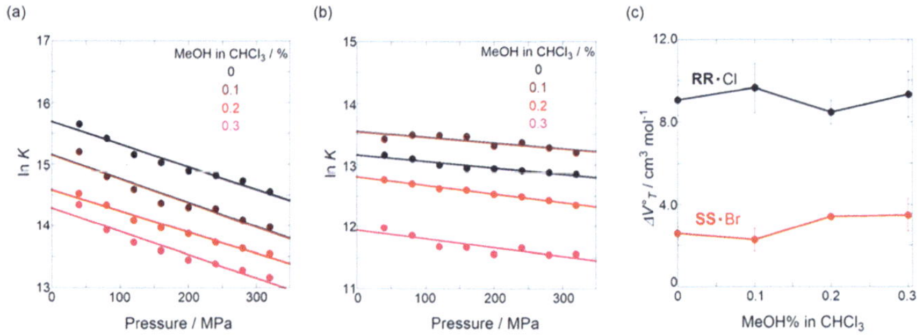

Fig. 5.17 Pressure dependence of the binding constant (K_{anion}) obtained from the supramolecular complexation of (**a**) **RR**·Cl and (**b**) **SS**·Br with **H** under hydrostatic pressures (40–320 MPa) at room temperature in chloroform containing 0–0.3% methanol. (**c**) Plots of ΔV°_{T} as a function of methanol content in chloroform. (Reproduced with permission. Copyright 2021 Royal Society of Chemistry)

strength points toward mutual specific solvation of both the complex and the free guest in the presence of methanol, rather than a simple dilution or bulk polarity effects. These observations indicate that subtle solvent microheterogeneity can modulate binding affinities independently of volumetric signatures.

Overall, the hydrostatic pressure and solvent environment cooperate to control the conformational distribution and optical response of the foldamer **H**, and modulate its molecular recognition toward both chiral and achiral ion pairs. Pressure-dependent spectroscopies reveal reversible shifts between folded and extended conformers in ground and excited states, as well as unexpected pressure-induced fluorescence quenching that arises from excited-state twisting and solvent reorganization. Furthermore, they quantify binding thermodynamics that are dominated by solvation-related volume changes. The ion-pair-driven chiral induction mechanism is particularly sensitive to pressure because the ion pair association modes differ in their solvated volumes, providing a route to pressure-tunable chirality transfer. These insights establish hydrostatic pressure as a versatile external control parameter for foldamer-based sensors and supramolecular systems whose function is governed by solvation and volumetric effects, and they suggest general design principles for next-generation pressure-responsive molecular devices.

5.4 Pressure-Controlled Chiral Recognition and Conformational Dynamics in a Flexible Naphthotube

Hydrostatic pressure was applied as an external, isotropic mechanical stimulus to investigate the dynamic control of chiral recognition by a water-soluble chiral naphthotube, denoted **7** (Fig. 5.18) (Motoori et al., 2024). This host features four naphthalene walls connected by a flexible methylene linker and possesses polar

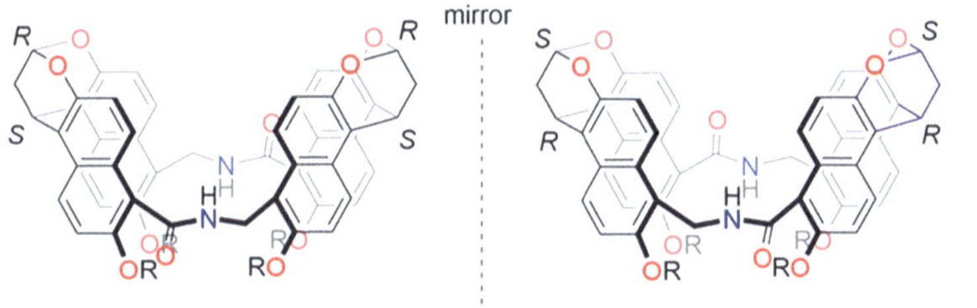

R^2,S^2-**7** S^2,R^2-**7**

R = $CH_2CO_2^-$ Na^+

Fig. 5.18 Chemical structure of compound **7**. (Reproduced with permission. Copyright 2024 American Chemical Society)

substituents that create a deeper hydrophobic cavity and a polar binding site (Chai et al., 2020; Yang et al., 2020). This cooperative structural motif differs fundamentally from those of cyclodextrins, calixarenes, cucurbiturils, and pillararenes. Spectroscopic and binding measurements under pressures up to 280 MPa, performed in aqueous media, probed ground- and excited-state optical properties, conformational dynamics, and quantitative thermodynamics of complexation with both hydrophilic and hydrophobic chiral guests. Enantiomeric guest pairs included the hydrophilic amino acids phenylalanine and tryptophan, as well as the hydrophobic styrene oxide. The primary objectives were to elucidate how the flexible naphthotube responds to hydrostatic pressurization and how solvation and desolvation processes influence enantioselectivity and apparent reaction volume changes, ΔV°.

Baseline chiroptical and photophysical characterizations at ambient pressure revealed that **7** exhibits a pronounced exciton-coupled CD in the long axis ^1B_b transition region, with anisotropy factor values of $g_{239\,\text{nm}} = -0.0024$ and $g_{226\,\text{nm}} = 0.0011$. These features are consistent with a left-handed arrangement of the four naphthalene walls according to exciton chirality theory (Fig. 5.19a). Fluorescence measurements revealed a relatively large Stokes shift (3280 cm^{-1}) and biexponential decay behavior, with a short component $\tau_1 \sim 0.4$ ns and a longer component $\tau_2 \sim 3.5$ ns (Fig. 5.19b, c). Wavelength-dependent abundance analysis indicated that the short-lived species dominate at longer emission wavelengths, implicating an intramolecular ground-state stacked species localized in the bis(naphthalene) cleft. The low quantum yield at ambient conditions ($\Phi_F \sim 0.07$) further supports the presence of nonradiative, stack-induced deactivation pathways. Structural evidence from single-crystal data corroborated the existence of a cleft acting as a hinge, underpinning a "butterfly"-type flapping motion that can be modulated by guest inclusion.

Guest-induced optical responses under ambient conditions exposed a dynamic flapping and fluorescence turn-on mechanism. Addition of the achiral model guest 1,4-dioxane produced negligible changes in absorption but induced a steady increase in fluorescence intensity (Fig. 5.20a, b), consistent with disruption of intramolecular stacking in the cleft upon guest inclusion. Nonlinear least-squares fitting of fluorescence titration data, assuming a 1:1 stoichiometry, yielded a binding constant

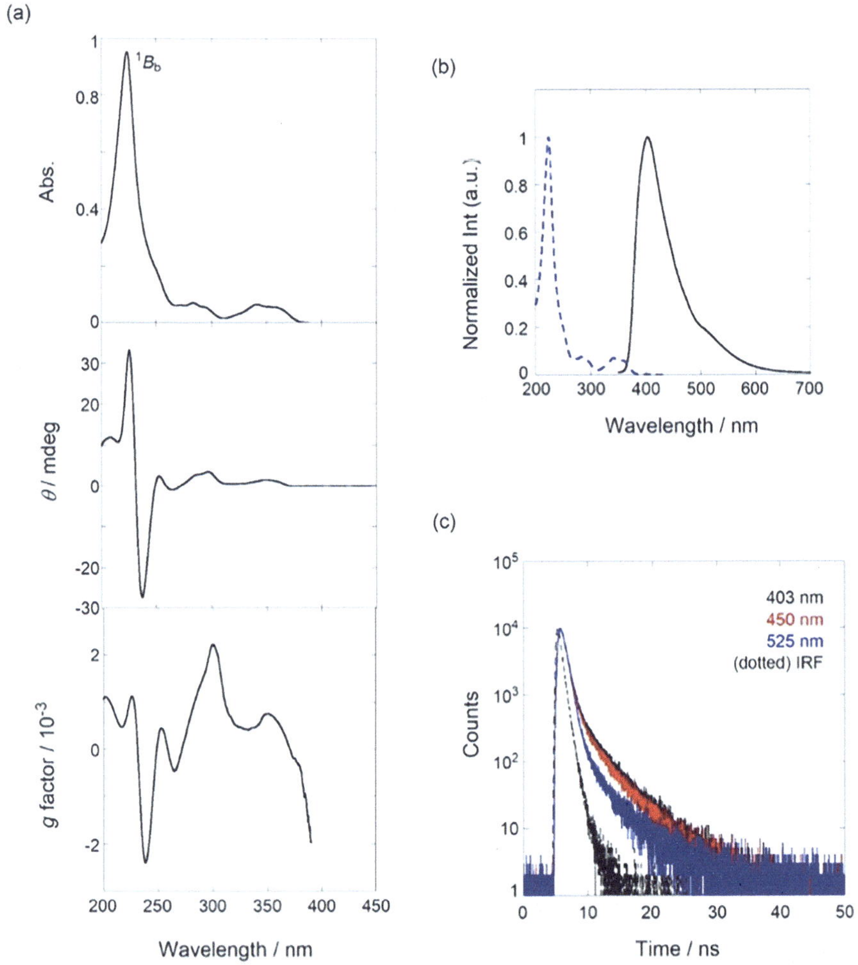

Fig. 5.19 (**a**) UV-vis (top), CD (middle), and *g*-factor spectra (bottom) of compound **7** (9.01 μM) in H_2O at 25 °C. (**b**) Fluorescence spectrum of **7** (9.01 μM) in H_2O at 25 °C (λ_{ex} = 300 nm, black solid line); the blue dotted line shows the normalized UV-vis spectrum. (**c**) Fluorescence decay profiles of **7** (9.01 μM) recorded at 403 nm (black), 450 nm (red), and 525 nm (blue) upon excitation at 280 nm at room temperature. All spectra were obtained using a 1-cm path-length cell. (Reproduced with permission. Copyright 2024 American Chemical Society)

$K \sim 2930\ M^{-1}$ (Fig. 5.20c). Concurrent time-resolved studies showed that τ_1 lengthened from 0.4 to 1.6 ns upon guest inclusion, whereas τ_2 remained nearly constant at 3.6–3.9 ns, indicating release of the stacked, nonemissive geometry and a corresponding turn-on of fluorescence (Fig. 5.20d). At host occupancies exceeding 99.9%, Φ_F increased to approximately 0.20, consistent with suppression of the short-lived, stack-related nonradiative channel and an induced-fit stabilization of the extended naphthalene geometry (Fig. 5.20e).

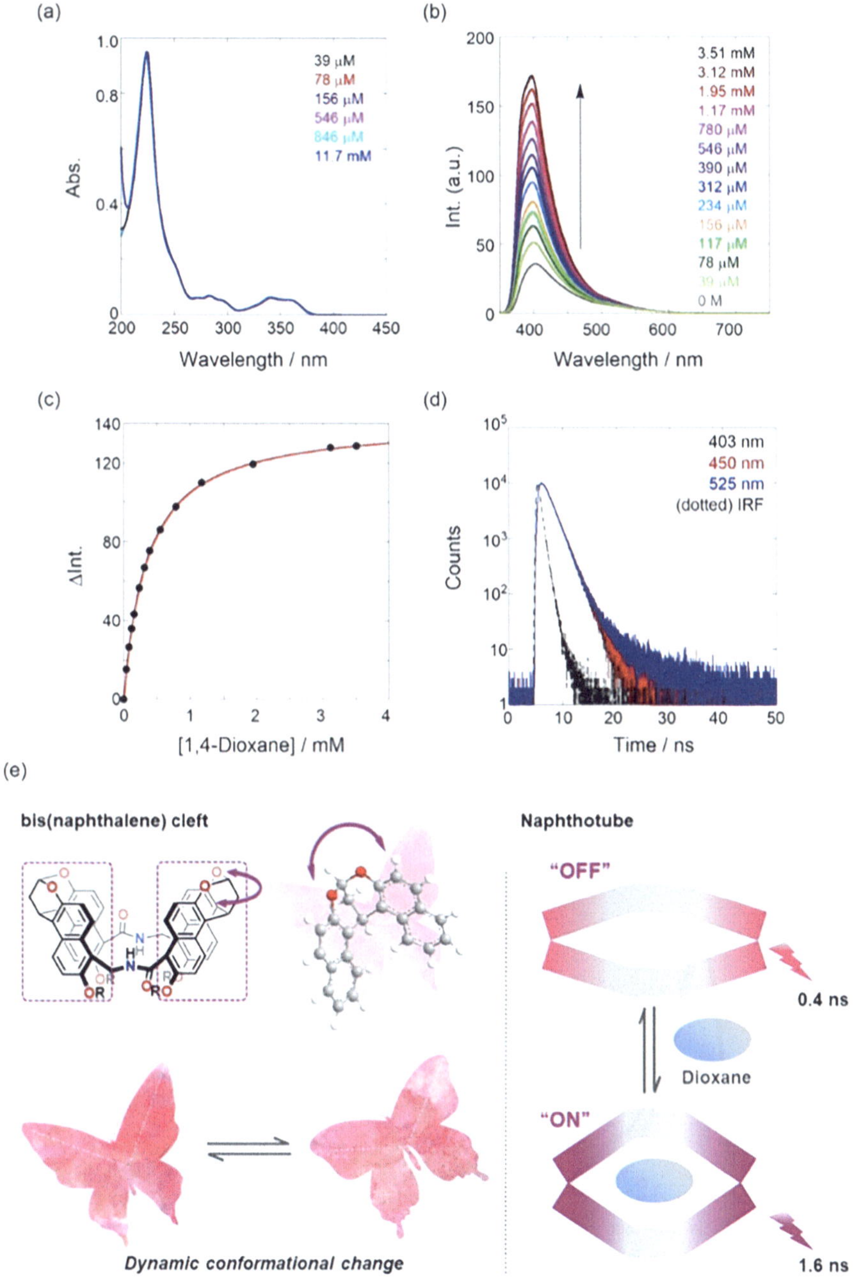

Fig. 5.20 (**a**) UV-vis and (**b**) fluorescence spectra (λ_{ex} = 300 nm) of **7** (9.01 μM) in H_2O at 25 °C upon gradual addition of 1,4-dioxane (0–3.51 mM, colored lines). (**c**) Nonlinear least-squares fitting based on a 1:1 binding model used to evaluate *K* between **7** and 1,4-dioxane. (**d**) Fluorescence decay profiles (λ_{ex} = 280 nm) of **7** (9.63 μM) in the absence and presence of 1,4-dioxane (5.56 mM), monitored at 403 nm (black), 450 nm (red), and 525 nm (blue) at room temperature. All spectra were obtained using a 1-cm path-length cell. (**e**) Schematic of the dynamic flapping motion (left) and fluorescence turn-on mechanism (right) of **7**. (Reproduced with permission. Copyright 2024 American Chemical Society)

Hydrostatic pressure spectroscopy in the absence of chiral guests demonstrated a reversible, pressure-dependent increase in absorbance attributable to effective concentration change under pressurization, together with an unexpected decrease in fluorescence intensity with increasing pressure (Fig. 5.21a, b). Normalized emission spectra revealed a relative growth of intramolecularly-stacked species in the longer wavelength region under pressurization (Fig. 5.21c). Time-resolved measurements across the pressure series showed subtle but systematic trends: the short lifetime τ_1 decreased from 0.42 to 0.35 ns while maintaining high relative abundance (~0.90), whereas τ_2 decreased from 3.2 to 2.3 ns, while remaining a minor component (Fig. 5.21d). These observations indicate that pressure promotes stacking within the bis(naphthalene) cleft, thereby enhancing nonradiative decay

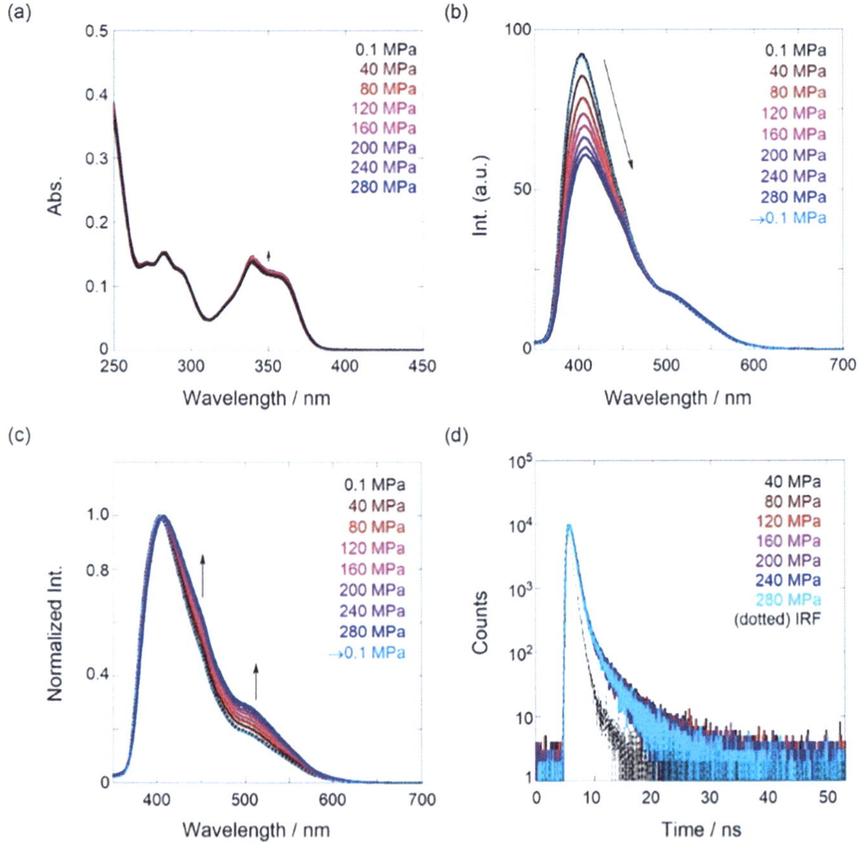

Fig. 5.21 Pressure-dependent (**a**) UV-vis spectra (90.1 μM), (**b**) fluorescence spectra (9.63 μM, λ_{ex} = 300 nm), (**c**) normalized fluorescence spectra, and (**d**) fluorescence decay profiles (87.9 μM, λ_{ex} = 280 nm, λ_{em} = 525 nm) of compound **7** in H_2O at room temperature under pressures ranging from 0.1 to 280 MPa. All measurements were conducted using a high-pressure optical cell. (Reproduced with permission. Copyright 2024 American Chemical Society)

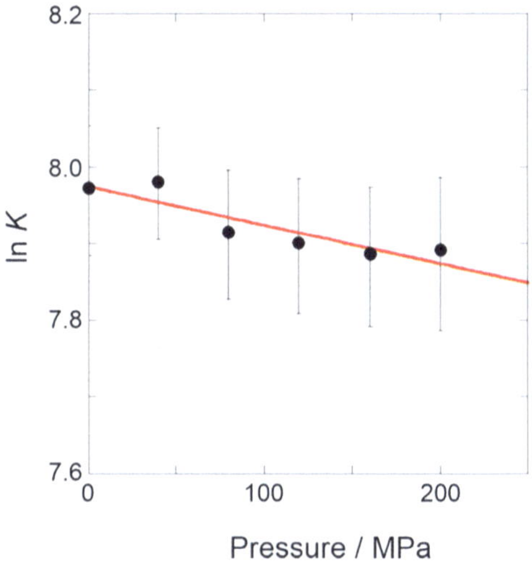

Fig. 5.22 Pressure dependence of K for 1,4-dioxane binding to **7** in H_2O under hydrostatic pressures (0.1 to 200 MPa) at room temperature ($r = 0.903$). (Reproduced with permission. Copyright 2024 American Chemical Society)

pathways. The recovery of the original spectrum upon depressurization further confirms that these effects are reversible and not destructive.

Quantitative analysis of pressure-dependent binding employed plots of ln$K$ versus pressure to derive apparent reaction volume changes (ΔV°) using the relation $(\partial \ln K/\partial P) = -\Delta V^\circ/RT$. For the achiral guest 1,4-dioxane, K values from 0.1 to 200 MPa followed a linear trend with pressure, yielding $\Delta V^\circ = +1.2 \pm 0.3$ cm^3 mol^{-1} (Fig. 5.22). Such a small positive value suggests that complex formation under pressurization is accompanied by a marginal increase in solvated molar volume, a result that can be rationalized by a preference for tighter stacking of the naphthalene walls rather than extensive extension. In other words, guest inclusion tends to reduce the accessible solvent volume modestly, and pressure slightly disfavors the complex relative to the uncompressed state.

Chiral recognition under pressure exhibited contrasting volumetric signatures depending on guest hydrophilicity. For hydrophilic amino acids, both phenylalanine (Phe) and tryptophan (Trp) delivered negative ΔV° values, consistent with a net reduction of solvated molar volume upon complexation under pressurization (Fig. 5.23c). Phenylalanine yielded ΔV° values of -0.9 ± 0.6 cm^3 mol^{-1} for D-Phe and $-$ 1.2 $\pm$ 0.2 cm^3 mol^{-1} for L-Phe (Fig. 5.23a), with enantioselectivities K_L/K_D ranging from about 1.3 to 1.6 across the pressure range. These small negative volumes were rationalized by preferential opening of the flexible naphthalene walls upon complexation, facilitating desolvation of water molecules around charged or polar functional groups on the guest, thereby producing a slight volumetric contraction. Trp exhibited markedly larger negative ΔV° values, -5.6 ± 0.5 cm^3 mol^{-1} for D-Trp and $-$ 7.0 $\pm$ 0.3 cm^3 mol^{-1} for L-Trp (Fig. 5.23b), while enantioselectivity remained modest between 1.2 and 1.4. The larger magnitude for Trp is attributable to the greater desolvation demands of its larger hydrophilic side chain, promoting more extensive opening of the naphthalene walls and a larger decrease in solvated volume upon binding.

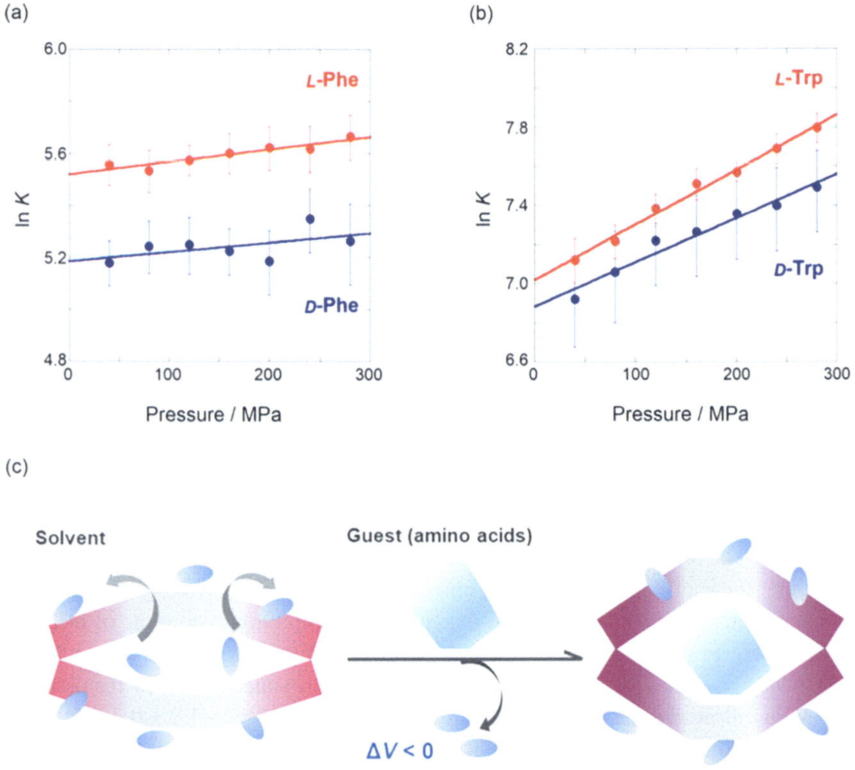

Fig. 5.23 Pressure-dependent variations in K for the complexation of (**a**) Phe ($r = 0.541$ for D, $r = 0.938$ for L) and (**b**) Trp ($r = 0.981$ for D, $r = 0.995$ for L) with **7** in H_2O under hydrostatic pressures ranging from 40 to 280 MPa at room temperature. (**c**) Schematic of the pressure-induced dynamic stretching of **7** triggered by guest inclusion. (Reproduced with permission. Copyright 2024 American Chemical Society)

The hydrophobic guest styrene oxide (guest **8**) exhibited a qualitatively different response. Binding constants were substantially larger for the *S* enantiomer, with enantioselectivities (K_S/K_R) between 5.9 and 7.6 across the pressure range, representing a notably stronger enantiodiscrimination than that observed for the amino acids. The $\Delta V°$ values for styrene oxide were positive, 3.5 ± 0.5 $cm^3\ mol^{-1}$ for *S*-**8** (Fig. 5.24a) and 1.5 ± 0.3 $cm^3\ mol^{-1}$ for *R*-**8** (Fig. 5.24b). The positive volumetric changes suggest that complexation traps both the hydrophobic guest and co-associated water molecules in a configuration that is volumetrically expanded relative to free solvated states, or that closure of the naphthalene walls around a strongly solvated hydrophobic guest increases the local solvated volume. The enhanced enantioselectivity is consistent with an induced-fit scenario in which the dynamic naphthotube tightly conforms to the size and shape of the hydrophobic guest, amplifying the diastereomeric binding energy differences relative to those observed for hydrophilic amino acids.

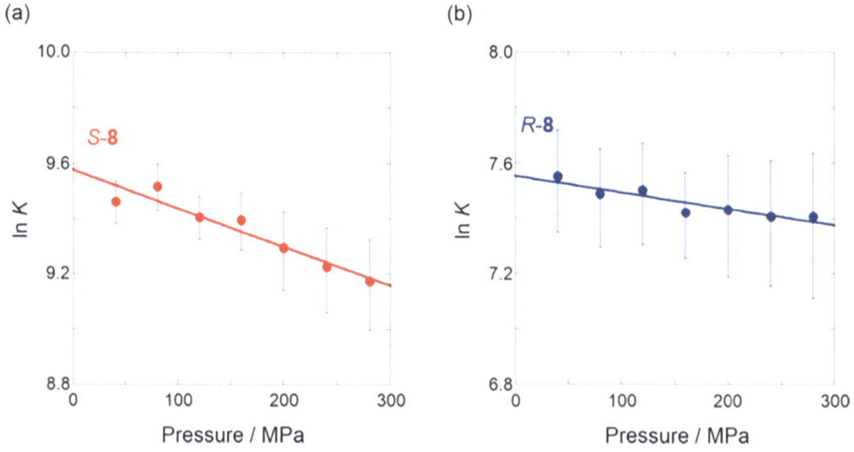

Fig. 5.24 Pressure-dependent changes in the binding constant (K) for the complexation of (**a**) *S*-**8** ($r = 0.955$) and (**b**) *R*-**8** ($r = 0.918$) with **7** in H_2O under hydrostatic pressures ranging from 40 to 280 MPa at room temperature. (Reproduced with permission. Copyright 2024 American Chemical Society)

In summary, the flexible methylene linker imparts dynamic motion to the naphthotube, enabling pressure-sensitive flapping and stacking equilibria in both ground and excited states. The sign and magnitude of ΔV° are governed by solvation and desolvation effects, determining whether hydrostatic pressure induced host opening or closing upon guest inclusion. Hydrophilic guests promote negative ΔV° and host expansion, whereas hydrophobic guests lead to positive ΔV°, host contraction, and enhanced enantioselectivity via induced fitting. These findings demonstrate that hydrostatic pressure serves as a tunable and reversible mechanical stimulus for controlling chiroptical and enantioselective recognition in water. This strategy provides a robust platform for developing pressure-responsive supramolecular systems, smart chemosensors, and molecular logic devices.

5.5 Multidimensional Control of Supramolecular Dimerization by Temperature and Hydrostatic Pressure

Supramolecular materials capable of responding to multiple external stimuli are central for the development of functional soft materials, because stimulus-driven structural and optical changes can be transduced into useful functions such as molecular memory, logic operations, and controlled release. Self-assembly, and in particular self-dimerization, provides a fundamental bottom-up strategy for constructing such responsive systems, because ordered supramolecular architectures can emerge from well-designed monomers through noncovalent interactions. Hydrostatic pressure, applied in solution as an isotropic, volumetric stimulus in the MPa range, remains comparatively underexplored despite its clear presence in thermodynamic expressions via a volume term and its potential relevance to mechanochemical and mechanobiological processes. A supramolecular system whose assembly state is both solvent dependent and kinetically accessible therefore offers attractive prospects for multidimensional control by combining pressure and temperature. A zinc phthalocyanine derivative bearing four pairs of 6,13-dihydro-6,13-diazapentacene (DHDAP) pillars and peripheral *n*-hexyl chains, denoted **9Zn** (Fig. 5.25a) (Saeki et al., 2019), fulfills these requirements because it forms an

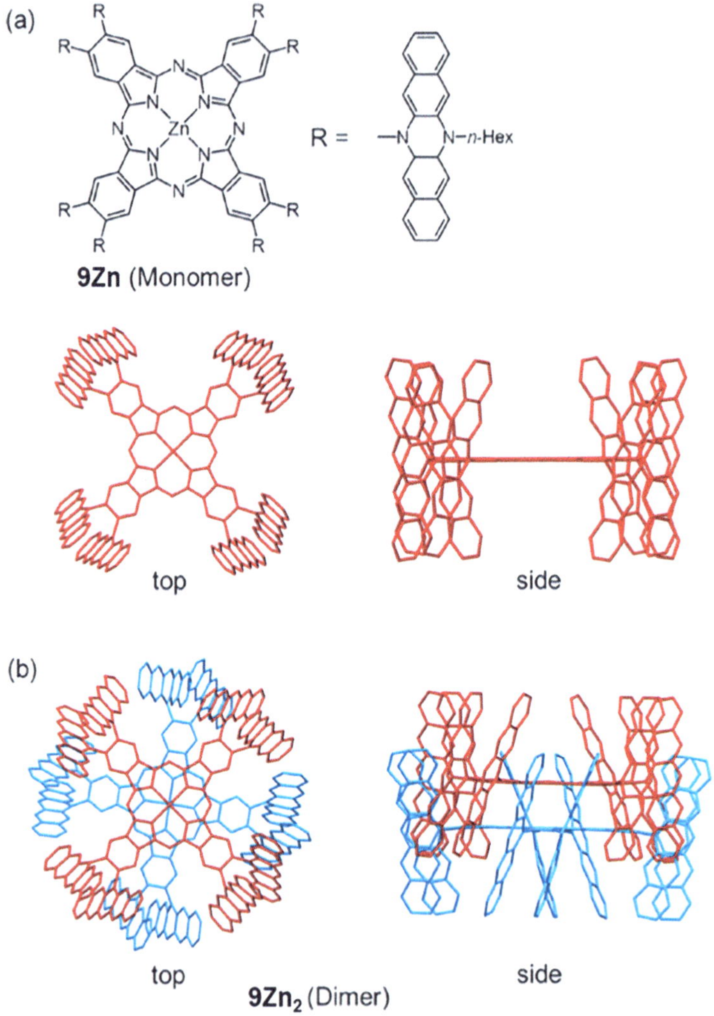

Fig. 5.25 Structures of the phthalocyanine derivatives: (**a**) monomeric form (**9Zn**) and (**b**) dimeric form (**9Zn$_2$**). *n*-Hexyl substituents are omitted for clarity. The 3D structures were generated from DFT-optimized geometries of model compounds without the *n*-hexyl groups. (Reproduced with permission. Copyright 2024 American Chemical Society)

H-type cofacial dimer, **$9Zn_2$** (Fig. 5.25b), through interdigitated pillars that engage like meshing gears, and its dimerization displays pronounced solvent sensitivity and kinetic barriers (Kinoshita et al., 2024).

Molecular modeling at the M06-2X/6-31G(d) level for H, C, N, and LANL2DZ for Zn, using truncated models without *n*-hexyl chains, revealed a strongly stabilized dimer. Two monomers interlock by filling each other's voids with DHDAP pillars, with nearly three of the five six-membered rings of each pillar inserted into counterpart voids. The computed stabilization energy upon dimerization was approximately 400 kJ mol^{-1}, comparable to covalent bond formation. The short Zn-Zn separation (3.315 Å) suggested a significant attractive interaction between monomers in the assembled state. Multiple short intermonomer contacts (<3.4 Å) supported the presence of interpillar π-π interactions in addition to the π-π stacking of the phthalocyanine planes. Structural characterization via mass spectrometry, one- and two-dimensional NMR for **9Zn** and **$9Zn_2$**, and ESR for the copper analogues confirmed a D_4-symmetric dimer with interdigitated pillars. The spatial constraints inherent to the dimer make trimer or higher oligomer formation unfavorable, consistent with experimental isolation of discrete monomer and dimer species.

Solvent effects were decisive in governing self-assembly. In THF, both monomer and dimer existed as stable, noninterconverting species even upon heating to 333 K, indicating large activation barriers for both association and dissociation due to steric demands associated with pillar interdigitation. In solvents such as DCM, toluene, and *o*-dichlorobenzene, the dimer formed as a metastable species that slowly dissociated to monomer, whereas in ethyl acetate (EA), the monomer converted dynamically to a metastable state that subsequently dimerized. These solvent-dependent behaviors therefore reflect how solvation cores surrounding the monomer influence desolvation energetics and steric accommodation during the dimerization transition.

Kinetic analysis focused on the dynamic dimerization in EA, where the process is experimentally tractable. Temperature-dependent UV/vis/NIR spectroscopy was conducted on 29 μM solutions by monitoring the phthalocyanine Q band, which appears at ~675 nm for the monomer and shifts hypsochromically to 652 nm upon formation of the *H*-type dimer. Time-dependent absorbance changes measured at 313–343 K displayed a monotonic decrease of the monomer band and emergence of the dimer band (Fig. 5.26a). The dimerization kinetics were well described by a second-order rate law, $d[\mathrm{M}]/dt = -k_{\mathrm{dim}}[\mathrm{M}]^2$, which upon integration produced reciprocal linear plots consistent with bimolecular association. Eyring analysis of the k_{dim} values obtained at four temperatures produced a straight line ($r = 0.999$), indicating a single dominant activated pathway over the tested range (Fig. 5.26b). The activation parameters determined from the slope and intercept were $\Delta H^{\ddagger} = 91.5$ kJ mol^{-1} and $\Delta S^{\ddagger} = +47.9$ J K^{-1} mol^{-1}, yielding $\Delta G^{\ddagger} = 77.2$ kJ mol^{-1} at 298 K. The positive activation entropy indicates that desolvation accompanies formation of the transition state, consistent with a model in which the solvent core around the monomer must be expelled to provide the effective free volume required for gear-like pillar engagement (Fig. 5.26c). The positive activation enthalpy reflects the energetic cost of such desolvation despite the large stabilization of the fully engaged dimer.

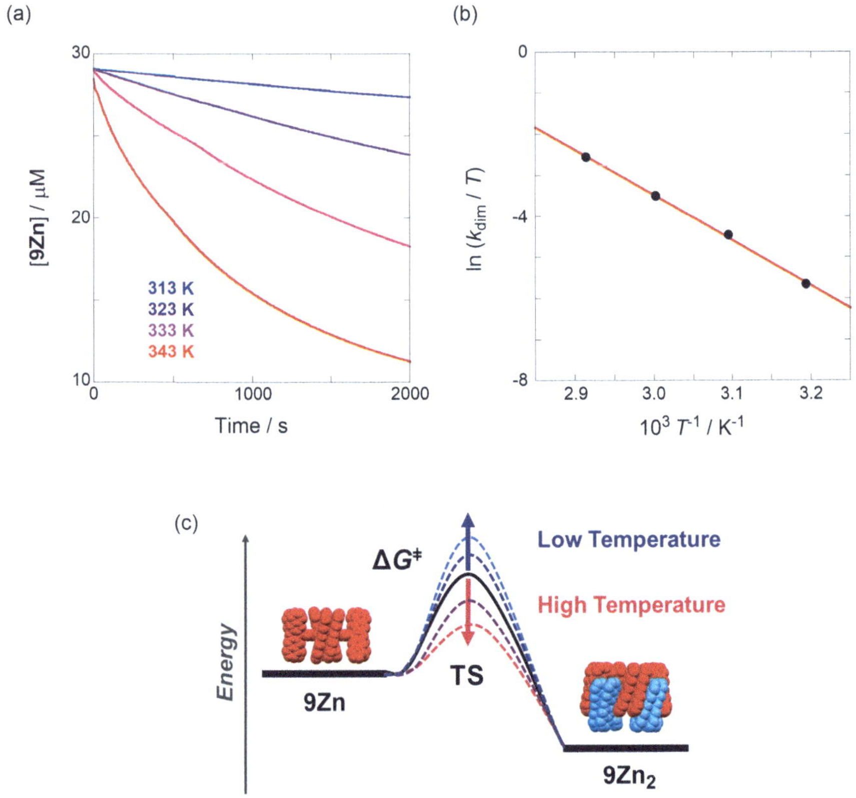

Fig. 5.26 (**a**) Time-dependent variation in the concentration of **9Zn** in EA measured at 0.1 MPa at 313, 323, 333, and 343 K (from blue to red). (**b**) Eyring plot for the dimerization of **9Zn** (29 µM) in EA at 0.1 MPa (r = 0.999). (**c**) Energy diagram illustrating the dimerization process of **9Zn** in EA under 0.1 MPa. (Reproduced with permission. Copyright 2024 American Chemical Society)

Hydrostatic pressure spectroscopy was conducted using a custom high-pressure cell capable of recording UV/vis/NIR spectra up to 320 MPa. Measurements in THF, toluene, DCM, and EA revealed pressure-induced bathochromic shifts and hyperchromic effects for both monomer and dimer. The bathochromic shifts were attributed to pressure-induced increases in solvent polarizability and density, which stabilize π* orbitals and reduce transition energies, while hyperchromism was attributed to the effective concentration increase upon solvent compression. The pressure-induced shift magnitudes for the Q band were modest relative to typical π → π* transitions, ranging from 0 to −0.39 cm^{-1} MPa^{-1} for **9Zn** and − 0.11 to −0.25 cm^{-1} MPa^{-1} for **$9Zn_2$**, reflecting the unique electronic structure of the phthalocyanine chromophore.

Crucially, hydrostatic pressurization exerted a kinetic control over dimer formation in EA at elevated temperatures. Time-resolved spectroscopic experiments at 333 K under ambient pressure displayed spontaneous conversion of the monomeric band at 675 nm to the dimer band near 649 nm over time (Fig. 5.27a). When identical experiments were performed at 100 MPa, the monomer band remained essentially unchanged during pressurization, indicating that pressure arrests dimerization (Fig. 5.27b). After maintaining 100 MPa for 60 min and then depressurizing to atmospheric pressure, the dimer band appeared within 20 min, demonstrating reversible kinetic control. The system therefore functions as a supramolecular gear with two distinct states, engaged and disengaged, whose activation toward the engaged state is promoted by temperature but inhibited by hydrostatic pressure (Fig. 5.27c).

Pressure-dependent kinetics were quantified by measuring k_{dim} at 20 MPa increments from 20 to 100 MPa at 333 K (Fig. 5.27d). A plot of $\ln k_{dim}$ versus pressure produced a straight line ($r = 0.972$), indicating that the nature of the transition state remained constant across the examined pressure range (Fig. 5.27e). From the slope, the activation volume $\Delta V^{\ddagger}$ was determined to be +62.0 $cm^3\ mol^{-1}$. This large positive activation volume is substantially greater than values reported for many pressure-responsive chemosensor systems, reflecting the formation of an expanded transition state in which the solvent must be expelled to create the space for gear engagement. The large positive $\Delta V^{\ddagger}$ is consistent with the previously observed positive $\Delta S^{\ddagger}$, further supporting desolvation as a key mechanistic element. The observed multidimensional control thus arises from competing effects: thermal activation facilitates overcoming the desolvation energy barrier, whereas increased pressure suppresses the solvent reorganization and spatial expansion required to reach the transition state.

In summary, a dynamic supramolecular gear based on Zn-coordinated phthalocyanine monomers with interdigitated DHDAP pillars exhibits solvent-dependent self-dimerization that can be controlled in a multidimensional fashion by temperature and hydrostatic pressure. Dimerization proceeds via a desolvation-driven transition state with large activation enthalpy and positive activation entropy, and hydrostatic pressure suppresses the assembly by imposing an unfavorable volumetric contribution described by a positive activation volume of +62.0 $cm^3\ mol^{-1}$. Spectroscopic monitoring indicates modest pressure-induced stabilization of π^* orbitals in compressed solvents and an increase in effective concentration upon pressurization, with bathochromic and hyperchromic signatures. The combination of detailed structural modeling, thermodynamic and kinetic analyses, and high-pressure spectroscopy establishes a clear mechanistic framework, in which desolvation and volumetric expansion in the transition state determine the accessibility of the engaged gear state, thereby enabling reversible, stimuli responsive control of supramolecular assembly.

→

Fig. 5.27 Time-dependent UV/vis/NIR absorption spectra of **9Zn** (32 μM) recorded in EA at 333 K under (**a**) 0.1 MPa (2 mm cell) and (**b**) 100 MPa (high-pressure cell). Spectra measured after standing for 0, 10, 20, 30, 40, 50, and 60 min are shown from black to light blue. The black dotted line in (**b**) represents the spectrum obtained after depressurization following 60 min at 100 MPa and 20 min at 0.1 MPa. (**c**) Proposed mechanism for the dimerization of **9Zn**. (**d**) Pressure-dependent changes in the concentration of **9Zn** observed in EA at 333 K under pressures of 20, 40, 60, 80, and 100 MPa (from black to purple). (**e**) Dependence of the dimerization rate constant (k_{dim}) on pressure in EA at 333 K ($r = 0.972$). (Reproduced with permission. Copyright 2024 American Chemical Society)

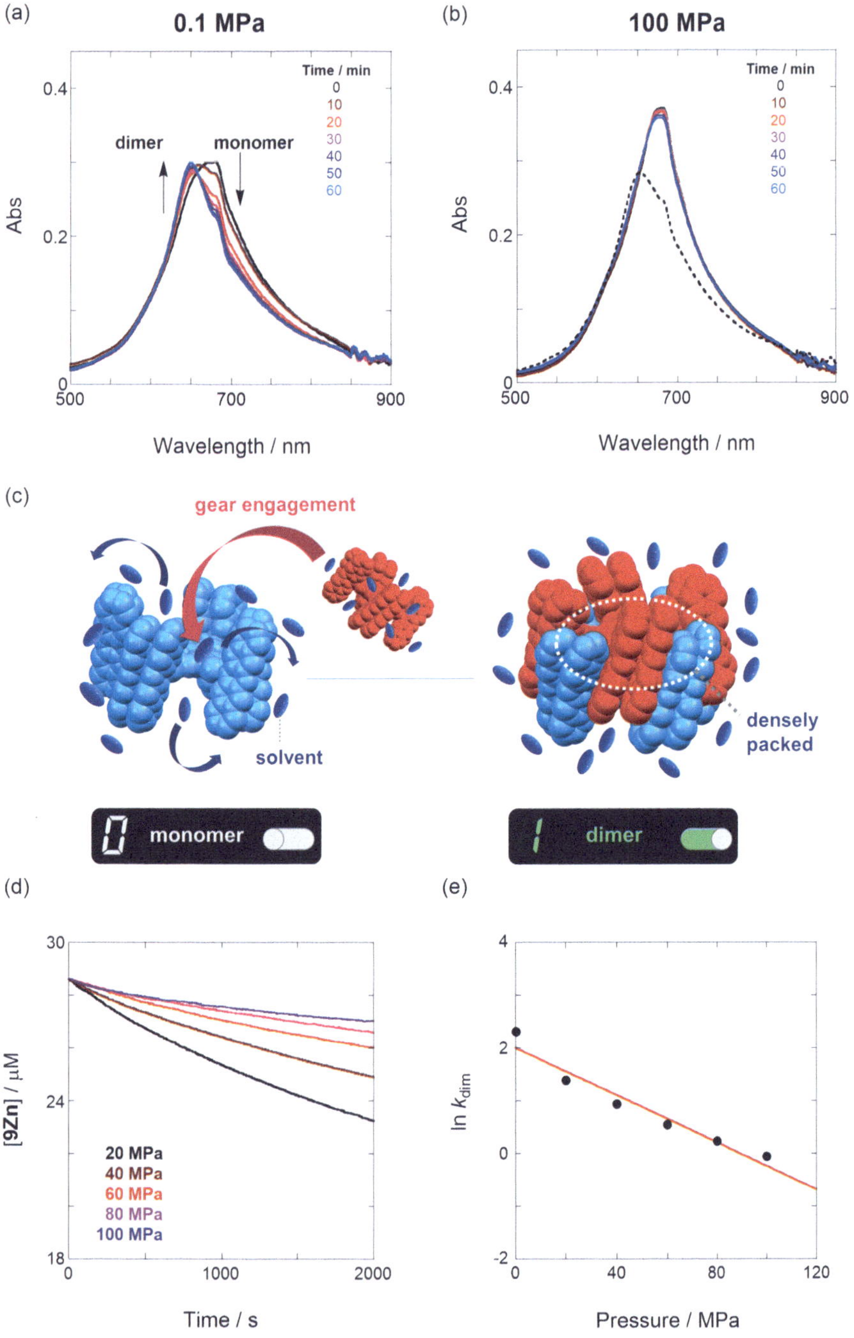
(a)
0.1 MPa
Time / min
0
10
20
30
40
50
60
dimer
monomer
Abs
0
0.2
0.4
500
700
900
Wavelength / nm
(b)
100 MPa
Time / min
0
10
20
30
40
50
60
Abs
0
0.2
0.4
500
700
900
Wavelength / nm
(c)
gear engagement
solvent
densely
packed
monomer
dimer
(d)
[9Zn] / μM
18
24
30
0
1000
2000
Time / s
20 MPa
40 MPa
60 MPa
80 MPa
100 MPa
(e)
ln k_dim
-2
0
2
4
0
40
80
120
Pressure / MPa

References

Chai, H., Chen, Z., Wang, S.-H., Quan, M., Yang, L.-P., Ke, H., & Jiang, W. (2020). Enantioselective recognition of neutral molecules in water by a pair of chiral biomimetic macrocyclic receptors. *CCS Chemistry, 2*, 440–452.

Haketa, Y., & Maeda, H. (2011). From helix to macrocycle: Anion-driven conformation control of π-conjugated acyclic oligopyrroles. *Chemistry - A European Journal, 17*, 1485–1492.

Haketa, Y., Bando, Y., Takaishi, K., Uchiyama, M., Muranaka, A., Naito, M., Shibaguchi, H., Kawai, T., & Maeda, H. (2012). Asymmetric induction in the preparation of helical receptor–anion complexes: Ion-pair formation with chiral cations. *Angewandte Chemie, International Edition, 51*, 7967–7971.

Harata, K., & Uedaira, H. (2006). The circular dichroism spectra of the β-cyclodextrin complex with naphthalene derivatives. *Bulletin of the Chemical Society of Japan, 48*, 375–378.

Heremans, K. (1982). High pressure effects on proteins and other biomolecules. *Annual Review of Biophysics, 11*, 1–21.

Ikeda, H., Nakamura, M., Ise, N., Oguma, N., Nakamura, A., Ikeda, T., Toda, F., & Ueno, A. (1996). Fluorescent cyclodextrins for molecule sensing: Fluorescent properties, NMR characterization, and inclusion phenomena of *N*-dansylleucine-modified cyclodextrins. *Journal of the American Chemical Society, 118*, 10980–10988.

Kinoshita, T., Haketa, Y., Maeda, H., & Fukuhara, G. (2021). Ground- and excited-state dynamic control of an anion receptor by hydrostatic pressure. *Chemical Science, 12*, 6691–6698.

Kinoshita, T., Sakamaki, D., & Fukuhara, G. (2024). Multidimensional dynamic control of supramolecular phthalocyanine gear: A self-assembly system responding to solvent, temperature, and hydrostatic pressure. *ACS Omega, 9*, 34719–34724.

Kodaka, M. (1993). A general rule for circular dichroism induced by a chiral macrocycle. *Journal of the American Chemical Society, 115*, 3702–3705.

Miyagawa, A., Yoneda, H., Mizuno, H., Numata, M., Okada, T., & Fukuhara, G. (2021). Hydrostatic-pressure-controlled molecular recognition: A steroid sensing case using modified cyclodextrin. *ChemPhotoChem, 5*, 118–122.

Motoori, J., Kinoshita, T., Chai, H., Li, M.-S., Wang, S.-M., Jiang, W., & Fukuhara, G. (2024). Dynamic control of chiral recognition in water-soluble naphthotubes induced by hydrostatic pressure. *ACS Nanoscience Au, 4*, 435–442.

Saeki, H., Sakamaki, D., Fujiwara, H., & Seki, S. (2019). Extreme multi-point van der Waals interactions: Isolable dimers of phthalocyanines substituted with pillar-like azaacenes. *Chemical Science, 10*, 8939–8945.

Sethy, R., Kumar, J., Métivier, R., Louis, M., Nakatani, K., Mecheri, N. M. T., Subhakumari, A., Thomas, K. G., Kawai, T., & Nakashima, T. (2017). Enantioselective light harvesting with perylenediimide guests on self-assembled chiral naphthalenediimide nanofibers. *Angewandte Chemie, International Edition, 56*, 15053–15057.

Sethy, R., Métivier, R., Brosseau, A., Kawai, T., & Nakashima, T. (2018). Impact of optical purity on the light harvesting property in supramolecular nanofibers. *The Journal of Physical Chemistry Letters, 9*, 4516–4521.

Yang, L.-P., Wang, X., Yao, H., & Jiang, W. (2020). Naphthotubes: Macrocyclic hosts with a biomimetic cavity feature. *Accounts of Chemical Research, 53*, 198–208.

Yonezawa, S., Sethy, R., Fukuhara, G., Kawai, T., & Nakashima, T. (2019). Pressure-dependent guest binding and release on a supramolecular polymer. *Chemical Communications, 55*, 5793–5796.

Chapter 6
Pressure-Responsive Polymer Chemosensors

Abstract This chapter discusses how polymer-based chemosensors convert hydrostatic pressure into optical signals. Decorating polymer backbones with emissive units creates ensembles in which chromophore spacing, chain conformation, and local solvation all contribute to the response. Hydrostatic pressure modulates intra- and interchain distances, reorganizes weak interactions, and alters free volumes around the emitters, thereby redistributing the balance among monomer-like, aggregate-like, and excimer-like emission channels in a predictable manner. In some designs, pressure controls competition between nonradiative decay and the formation of emissive species, enabling dual emission that can be read out ratiometrically. In others, conformational adjustment of the polymer scaffold under pressure reversibly shifts the population of emissive states, allowing quantitative, reversible pressure sensing even in viscous or buffered media. Time-resolved and chiroptical measurements clarify the underlying excited-state manifolds and reveal how chain flexibility, chromophore density, and steric configuration determine the form and dynamic range of the pressure response. Together, these results establish general design rules for mechanoresponsive polymer chemosensors, in which volumetric compression is translated into robust ratiometric and chiral optical outputs that are suitable for solution-phase mechanosensing and mechanobiological interrogation.

Keywords Hydrostatic pressure · Polymer chemosensors · Aggregation-induced emission · Excimer

6.1 Pressure-Tunable Dual Emission in AIE-Active Polymer Systems

The development of dual-emission materials whose relative emission intensities can be controlled ratiometrically through various external stimuli, such as temperature, solvent environment, and pH, has attracted increasing attention (Cui et al., 2016; Haidekker & Theodorakis, 2016; Hawes & Gunnlaugsson, 2017; Lee et al., 2015;

T. Hamachi, G. Fukuhara, *Hydrostatic Pressure Chemistry*,
https://doi.org/10.1007/978-981-95-7461-2_6

McLaurin et al., 2013; Park et al., 2020; Yang et al., 2019). Dual emission refers to the phenomenon in which a material exhibits two distinct fluorescence bands. This feature allows quantitative detection of environmental changes by monitoring the variation in relative intensity ratios between the two bands. Such ratiometric approaches are highly valuable because they reduce the influence of extrinsic factors, such as fluctuations in excitation intensity or sample concentration, providing more reliable and reproducible signals compared to single-channel fluorescence measurements. Traditionally, the design of dual-emission materials has relied on specific excited-state processes, including excimer or exciplex formation (Sarma & Wong, 2018; Karuppannan & Chambron, 2011), TICT (Grabowski et al., 2003; Sasaki et al., 2016), excited-state intramolecular proton transfer (ESIPT) (Padalkar & Seki, 2016; Sedgwick et al., 2018), and TADF (Yanai & Kimizuka, 2017; Yang et al., 2017). The emission characteristics arising from these excited-state processes are often highly sensitive to solvent polarity, viscosity, and temperature. Furthermore, recent studies suggested that such emission features can also be influenced by hydrostatic pressure, which represents an isotropic, uniform mechanical stimulus that can alter molecular packing and conformational dynamics.

Among ratiometric fluorescence strategies, aggregation-induced emission (AIE) represents a promising platform (Luo et al., 2001; Tong et al., 2006; Mei et al., 2015). In AIE systems, the transition between monomeric and aggregated states can be leveraged to achieve ratiometric sensing. Tetraphenylethylene (TPE) is a prototypical AIE molecule that exhibits minimal fluorescence in dilute solution due to free rotation of its peripheral phenyl groups. In contrast, when aggregated, the rotational motions are restricted, leading to strong fluorescence. However, applying hydrostatic pressure as a stimulus introduces a unique challenge. While pressure is isotropic and uniform, it can also induce dissociation of preformed aggregates through volume reduction, leading to fluorescence quenching rather than enhancement. This phenomenon implies that conventional AIE systems may face difficulties in achieving ratiometric fluorescence responses under hydrostatic pressure, because the balance between monomeric and aggregated states can be disrupted.

To overcome this limitation, photochemical reactions were incorporated into AIE-based polymer systems to achieve ratiometric control. Specifically, TPE undergoes a photochemical cyclization and oxidation reaction upon irradiation, forming 9,10-diphenylphenanthrene (DPP, Fig. 6.1a). DPP fluoresces primarily in the 350–450 nm range, which is well-separated from the typical AIE emission band of TPE (approximately 400–600 nm). This clear spectral separation provides an opportunity to quantitatively monitor the relative contributions of TPE and DPP fluorescence, enabling ratiometric measurements. In this section, TPE units were incorporated into polyethylene polymer backbones to enhance the efficiency of TPE-to-DPP photoconversion under photoexcitation and to investigate how hydrostatic pressure modulates this photochemical pathway (Nakasha & Fukuhara, 2020).

A series of polyethylene derivatives bearing branched TPE units were synthesized with varying degrees of substitution (DS) of TPE along the polymer chain. To ensure solubility in aqueous systems, trimethylamino functional groups were introduced, allowing measurements in a DMSO-water (1:9) mixed solvent. Three

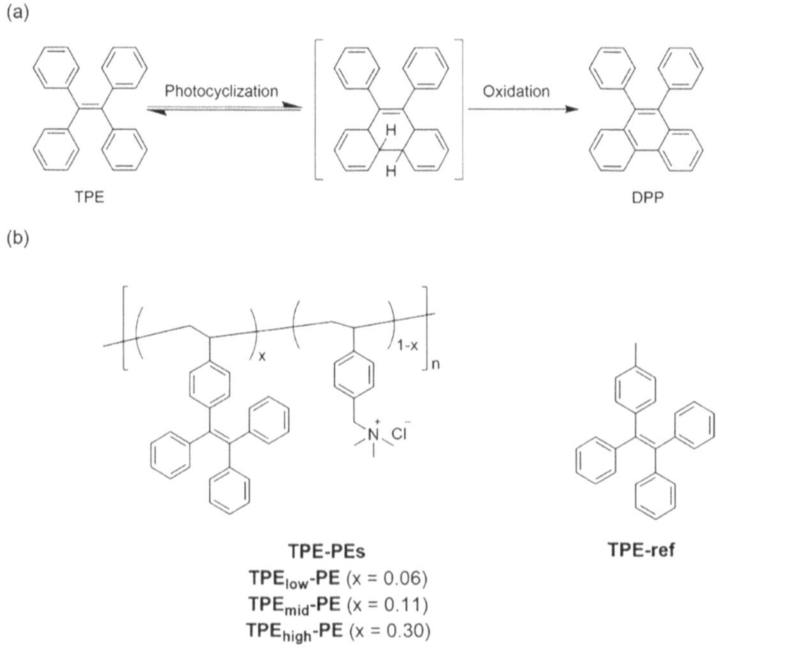

Fig. 6.1 (**a**) Photocyclization of TPE and subsequent photo-oxidation. (**b**) Molecular structures of **TPE-PEs** and **TPE-ref**. (Reproduced with permission. Copyright 2020 American Chemical Society)

polymers with low (**TPE$_{low}$-PE**), medium (**TPE$_{mid}$-PE**), and high (**TPE$_{high}$-PE**) substitution degrees were prepared (Fig. 6.1b). A small-molecule TPE reference compound (**TPE-ref**) was also used as a control to evaluate the role of polymer incorporation in promoting the photochemical reaction. Photophysical properties were characterized under ambient conditions, and ratiometric fluorescence responses were evaluated under hydrostatic pressure.

To determine whether the observed fluorescence response originated from intrachain processes rather than interchain aggregation, UV-vis absorption and dynamic light scattering (DLS) measurements were performed. For **TPE$_{low}$-PE**, UV-vis absorption followed the Beer-Lambert law over a range of concentrations, indicating negligible aggregation. DLS measurements confirmed small particle sizes of 1–3 nm. Even for **TPE$_{high}$-PE**, particle sizes remained below 12 nm, indicating the absence of large micelle-like aggregates (typically ~40 nm according to previous studies). This evidence confirms that fluorescence responses arise primarily from intrachain phenomena rather than interchain aggregation.

Fluorescence spectra were then recorded at room temperature under atmospheric pressure. **TPE$_{low}$-PE** exhibited the characteristic AIE band (400–600 nm), accompanied by a minor shoulder at approximately 380 nm. Upon repeated irradiation (58 μW cm^{-2}/run), the shoulder evolved into a structured band characteristic of DPP fluorescence, and an isoemissive point appeared near 438 nm (Fig. 6.2a). The resulting spectrum matched that of the monomer DPP reference, with only minor shifts attributed to solvent effects. Fluorescence lifetime measurements further confirmed that the fast component observed at 380 nm (approximately 0.1 ns) corresponds to DPP formation, whereas the longer-lived emission at 510 nm (1.5–1.8 ns) arises from AIE.

In contrast, **TPE-ref** did not exhibit DPP formation under identical conditions, instead showing only fluorescence quenching due to photobleaching (Fig. 6.2b). Such a finding highlights the critical role of polymer incorporation in facilitating the TPE-to-DPP photochemical conversion. The partial restriction of phenyl-ring rotation upon attachment to the polyethylene backbone likely facilitates photochemical cyclization, making polymer incorporation a decisive factor.

To quantitatively assess the balance between AIE decrease and DPP formation, a photoresponse factor was defined. The slope of the DPP emission increase (380 nm)

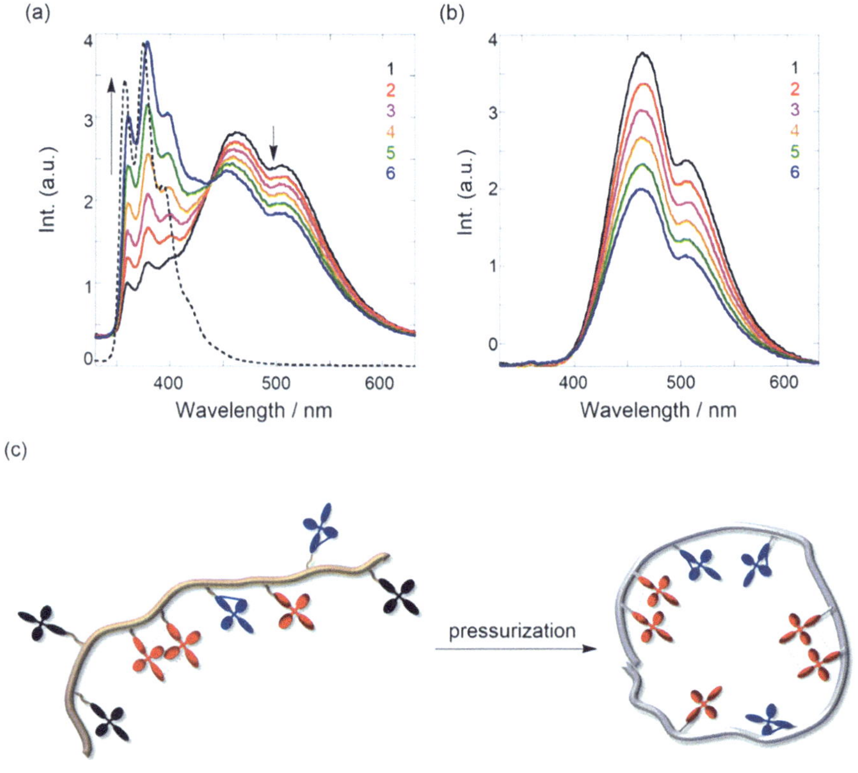

Fig. 6.2 Fluorescence spectra recorded at 320 nm excitation in DMSO-H_2O (1:9, v/v) solutions of (**a**) **TPE$_{low}$-PE** (9.8 μM) and (**b**) **TPE-ref** (9.8 μM) at 0.1 MPa and room temperature in a high-pressure cell, obtained during the first (black), second (red), third (pink), fourth (orange), fifth (green), and sixth (blue) irradiation cycles. (**c**) Schematic of the proposed mechanism for the pressure-induced ratiometric responses of **TPE-PEs**, where the black clover denotes ground-state TPE, the red clover denotes excited-state TPE, and the blue clover denotes excited-state DPP. (Reproduced with permission. Copyright 2020 American Chemical Society)

was denoted as α, the slope of the AIE emission change (504 nm) was denoted as β, and the ratio $\gamma = \alpha/\beta$ was calculated. Values of γ greater than 1 indicate dominant DPP formation, while γ less than 1 reflects the predominance of AIE reduction. For **TPE_{low}-PE**, γ ~5.2 under atmospheric pressure, indicating that DPP formation is strongly favored. **TPE_{mid}-PE** exhibited γ ~2.0, suggesting a moderate preference for DPP formation, whereas **TPE_{high}-PE** showed γ ~0.1, indicating that AIE reduction dominates. The reduction in α with increasing TPE substitution is attributed to steric hindrance around the branched TPE units, which limits oxygen access to the excited TPE molecules and thereby suppresses DPP formation.

Given the flexibility of the polyethylene backbone, these polymers are susceptible to structural changes under hydrostatic pressure, which can modulate their ratiometric fluorescence response. For **TPE_{low}-PE**, both α and β increased with increasing pressure, and the DPP emission band grew rapidly. The γ value remained high (approximately 4–5), demonstrating that pressure further promotes DPP formation. Conversely, **TPE_{high}-PE** displayed small α and large β, with γ being consistently below 1, indicating that pressure preferentially enhances AIE reduction rather than DPP formation. **TPE_{mid}-PE** exhibited intermediate behavior, where pressure effects on α and β were balanced, resulting in minimal net change in γ.

This pressure dependence can be explained mechanistically. In low-substitution polymers, the TPE units remain relatively isolated even when the polymer chains contract under pressure. This accessibility facilitates oxygen diffusion and photochemical conversion to DPP, resulting in substantial pressure-dependent increases in α (Fig. 6.2c). In high-substitution polymers, pressurization densely packs TPE units, hindering oxygen diffusion and suppressing DPP formation, while β changes significantly due to enhanced AIE quenching. Medium-substitution polymers display a balance between these opposing effects, producing intermediate responses.

In conclusion, this study demonstrates that ratiometric dual emission can be achieved in AIE-based polymer systems through the application of hydrostatic pressure. Specifically, TPE-modified polyethylene polymers allow the formation of two spectrally well-separated fluorescence channels via photoconversion of TPE to DPP under irradiation. This photoconversion process is highly sensitive to hydrostatic pressure, enabling precise tuning of the balance between the AIE-derived fluorescence decrease and DPP-derived fluorescence increase. Consequently, ratiometric fluorescence responses reflecting pressure changes can be realized. Additionally, the degree of TPE substitution was identified as a key factor governing the pressure response: low-substitution polymers promote DPP formation under pressure, while high-substitution polymers favor AIE reduction. These findings provide a new design principle for AIE-based polymer materials that exhibit ratiometric fluorescence responses to hydrostatic pressure, offering a valuable framework for the development of next-generation smart photofunctional materials with finely tunable emission properties.

6.2 Pressure-Modulated Excimer Formation in Pyrene–Peptide Conjugates

The previous section described the implementation of a ratiometric fluorescence readout, with the signal being irreversible due to reliance on a photochemical reaction. To realize a reversible ratiometric sensor, a design employing pyrene dyes was therefore proposed. Pyrene readily forms excimers under high chromophore concentrations via association of an excited-state molecule with a ground-state monomer. Because excimer emission appears at longer wavelengths than monomer fluorescence, pyrene-based systems possess intrinsic potential for ratiometric fluorescence readout. This section examines how the placement of pyrene chromophores on a peptide scaffold (Nishikawa et al., 2016; Mimura et al., 2018), together with the presence and positional arrangement of bulky piperidine auxiliaries, modulates the pressure-dependent photophysical behavior of the conjugates (Mizuno et al., 2020). Peptide oligomers with varying numbers of methylene spacers between pyrene units, together with derivatives bearing piperidine substituents, were synthesized. Absorption, fluorescence, and CD spectra were measured to determine how chromophore spacing and steric substituents influence optical and chiroptical properties.

Concentration-dependent measurements were performed for a representative conjugate, **L2C4** (Fig. 6.3a), over 44.4–298 μM, revealing a linear dependence of the fluorescence maxima at ~379 nm on concentration and no evidence of self-quenching up to ~227 μM. From these observations it was concluded that the pressure-induced spectral changes observed in this study originate predominantly from intrachain interactions among pyrenyl units appended to the peptide scaffold, rather than from intermolecular aggregation in solution at the examined concentrations.

Chromophore location exerted a decisive influence on the pressure response. Two representative constructs were compared: **L2C1**, with closely positioned

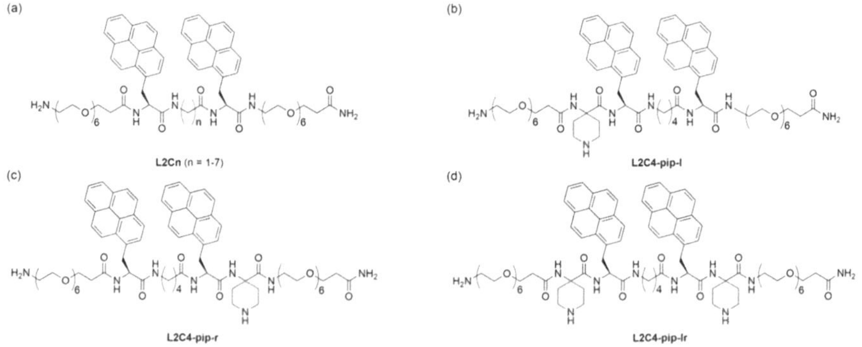

Fig. 6.3 Molecular structures of (**a**) **L2Cn** (n = 1–7), (**b**) **L2C4–pip–l**, (**c**) **L2C4–pip–r**, and (**d**) **L2C4–pip–lr**. (Reproduced with permission. Copyright 2020 John Wiley & Sons)

pyrenyl units, and **L2C4**, with more widely separated chromophores. The absorption spectra of **L2C4** exhibited a steady bathochromic shift of $-1.09\ cm^{-1}\ MPa^{-1}$ around 350 nm, together with increasing absorbance under pressure (Fig. 6.4b). The bathochromic shift was attributed primarily to pressure-induced changes in solvent polarizability, while the hyperchromic behavior was rationalized by an effective concentration increase under pressurization. CD and anisotropy signals for **L2C4** showed modest but progressive increases in the 280–300 nm region, indicating a pressure-driven chiral conformational change.

Conversely, **L2C1** displayed an anomalous absorption decrease between 0.1 and 80 MPa, before converging toward a common pressure-dependent red shift at higher pressures, with a slope of approximately $-1.11\ cm^{-1}\ MPa^{-1}$ above 120 MPa (Fig. 6.4a). The initial absorbance loss was interpreted as evidence for strong ground-state stacking of pyrenyl chromophores in **L2C1**. Consistent with this interpretation, the CD spectra for **L2C1** developed stepwise negative Cotton effects in the S_1 region of the pyrenyl band, supporting the idea of pressure-induced changes in chiral stacking geometry.

The fluorescence behavior under pressure was classifiable into three spectral regimes that were assigned to distinct emissive species. Short-wavelength emission (approximately 350–420 nm) was assigned to monomer fluorescence (Fig. 6.5). Emission in the mid-wavelength region (approximately 420–500 nm) was attributed to radiative decay from ground-state dimers. Long-wavelength emission (approximately 500–650 nm) was ascribed to excimer fluorescence arising from excited-state π-stacking. These assignments were supported by comparisons between excitation spectra and steady-state absorption, as well as by lifetime measurements. For **L2C1**, the short-wavelength emission exhibited a 20–25 ns lifetime consistent with monomer fluorescence; the mid-wavelength band displayed a much shorter lifetime (0.9–1.3 ns), indicative of radiative decay from a ground-state dimer; and the long-wavelength emission showed a lifetime on the order of tens of nanoseconds, consistent with excimer fluorescence.

A particularly revealing observation was the presence of a rise component in the time-resolved decay profiles recorded in the excimer spectral region for **L2C1**. This rise component had a characteristic time of roughly 10–13 ns and was interpreted as formation kinetics of a frustrated precursor excimer (Fukuhara et al., 2015), sometimes described as a second excimer state that precedes the fully overlapped sandwich excimer. Global analysis of abundances revealed that pressure gradually reduced the relative abundance of the monomer species while increasing the presence of ground-state dimer and excimer species. Quantitatively, the ratio of the fully overlapped excimer amplitude to the negative amplitude associated with the rise component ($A_e/(-A_f)$) increased from ~3.4 to ~5.1 with increasing pressure, consistent with a pressure-driven conversion of the frustrated precursor into the fully overlapped excimer.

Excitation spectra provided additional mechanistic insight. At ambient pressure, excitation spectra monitored at longer emission wavelengths overlapped with absorption bands near 350 nm, consistent with excimer formation from a sandwich-type stacking geometry in the excited state (Fig. 6.6a). Upon pressurization to

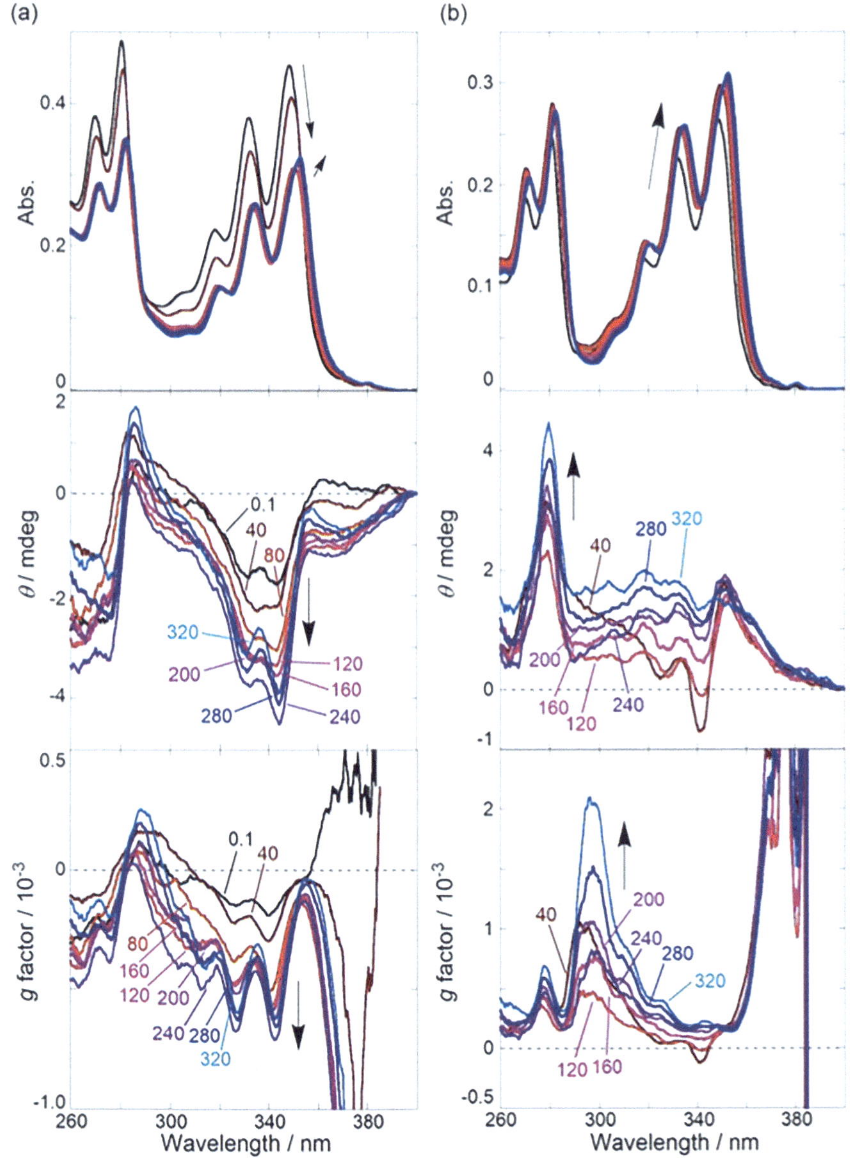

Fig. 6.4 UV-vis (top), CD (middle), and anisotropy ($g = \Delta\varepsilon/\varepsilon$, bottom) spectra of (**a**) **L2C1** (57.0 μM) and (**b**) **L2C4** (38.2 μM) in $CHCl_3$ at room temperature, recorded in a high-pressure cell at 0.1–320 MPa (black to sky blue lines). CD and g factor spectra of **L2C4** at 0.1 and 80 MPa were omitted due to baseline issues. (Reproduced with permission. Copyright 2020 John Wiley & Sons)

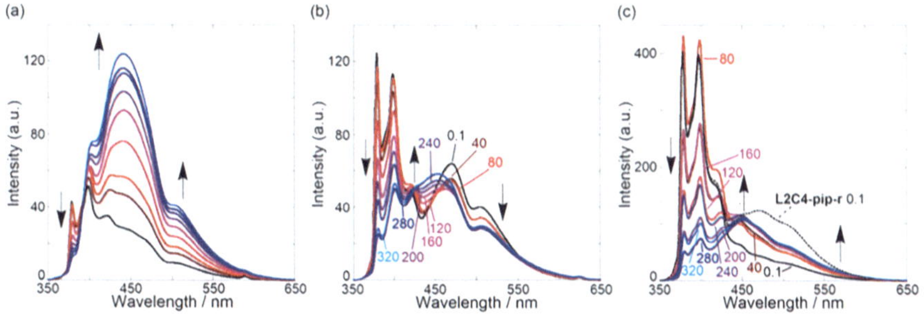

Fig. 6.5 Fluorescence spectra of (**a**) **L2C1** (57.0 μM, λ_{ex} = 295 nm), (**b**) **L2C4** (38.2 μM, λ_{ex} = 313 nm), and (**c**) **L2C4–pip–Ir** (128 μM, λ_{ex} = 310 nm) in $CHCl_3$ at room temperature, recorded in a high-pressure cell at 0.1–320 MPa (black to sky blue lines). In (**c**), the dotted black line represents the spectrum of **L2C4–pip–r** at 0.1 MPa, normalized at 378 nm. (Reproduced with permission. Copyright 2020 John Wiley & Sons)

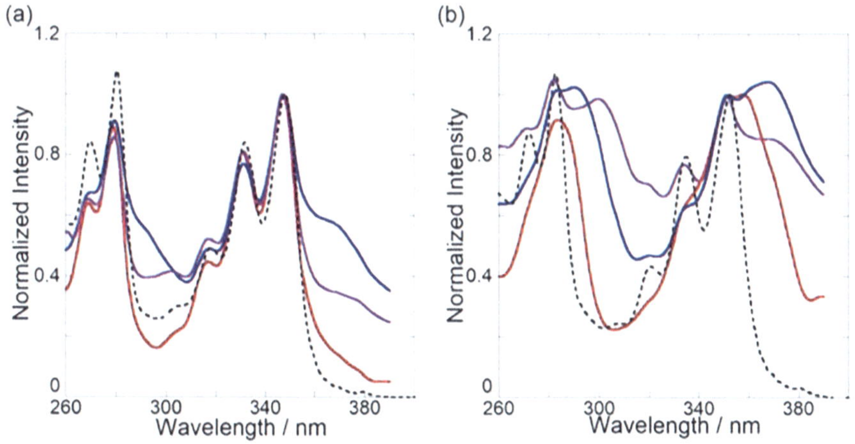

Fig. 6.6 UV-vis absorption (dotted) and excitation spectra (solid, monitored at 400 nm in red, 450 nm in blue, and 520 nm in purple) of **L2C1** (57.0 μM) in $CHCl_3$ at room temperature, recorded in a high-pressure cell at (**a**) 0.1 MPa and (**b**) 320 MPa. (Reproduced with permission. Copyright 2020 John Wiley & Sons)

320 MPa, non-superimposable excitation features emerged near 300 and 380 nm by shifting monitoring wavelengths, strongly indicating the development of ground-state dimer species under pressurization (Fig. 6.6b). Together, steady-state and time-resolved spectroscopies indicate that, for closely-spaced chromophores, hydrostatic pressure enhances ground-state stacking and promotes a sequence in which ground-state dimer formation and a frustrated excimer precursor give way to the formation of the fully overlapped excimer in the excited state.

In contrast, **L2C4**—where chromophores are more widely separated on the scaffold—showed suppressed formation of both ground-state dimers and excimers under pressure. The fluorescence intensities decreased in the short- and long-wavelength regions, while only modest increases were observed in the mid-wavelength region. Lifetime and excitation analyses nevertheless supported the same three emissive classes, but the pressure-dependent trends were far less pronounced. Such results demonstrate that the distance between chromophores on the peptide backbone critically governs the propensity for pressure-induced aggregation-like structures and the corresponding emissive outcomes. A systematic comparison across the **L2Cn** series with varying inter-chromophoric separations confirmed that dimer and excimer populations decline with increasing chromophore separation, establishing a clear structure–property relation.

The influence of bulky substituents on the peptide scaffold was examined using piperidine-substituted derivatives of **L2C4**. Three variants carrying piperidine at different positions were characterized. Their absorption bands shifted bathochromically under pressure with slopes of approximately −0.80 to −0.85 cm^{-1} MPa^{-1}, comparable to the parent **L2C4**. However, the sign and magnitude of the induced Cotton effects varied among these derivatives, indicating that piperidine placement exerts a decisive control over pressure-sensitive chiroptical properties.

When a single piperidine moiety was installed, only modest changes in the formation of ground-state dimer and excimer species were observed under pressure, indicating that one bulky auxiliary does not strongly frustrate chromophore stacking. In marked contrast, the derivative bearing two piperidine units on opposite sides of the scaffold exhibited a substantially altered fluorescence spectrum at ambient pressure, with reduced intensity in the mid- and long-wavelength regions. Time-resolved decay analysis for this doubly substituted conjugate revealed four exponential components but importantly no discernible rise component in the excimer region. The short and intermediate lifetimes (~0.8–5.1 ns) were attributed to two distinct types of ground-state dimers, whereas two long-lived components (~36–44 ns) were assigned to monomer and excimer emission in different spectral windows. Crucially, pressure increased the abundance of the two ground-state dimer contributions while suppressing monomer and excimer emission, demonstrating that the presence of two bulky piperidine auxiliaries sterically hinders the formation of the frustrated excited-state precursor and thereby modifies the pressure-response pathway.

In conclusion, hydrostatic pressure modulates both ground- and excited-state organization of peptide-linked pyrene arrays in a manner that is strongly dependent on interchromophoric distance and steric auxiliaries. Beyond solvent-polarizability-driven spectral shifts, pressure enhances effective chromophore proximity and promotes structural reorganization when sufficient spatial flexibility is available. For closely-spaced chromophores, pressure induces ground-state dimer formation and facilitates the conversion of a frustrated excimer precursor into a fully overlapped excimer in the excited state, whereas increased interchromophoric separation or steric hindrance suppresses these pathways. Chiroptical changes revealed by CD and anisotropy measurements confirm that pressure-induced conformational adjustments of the peptide backbone underlie the observed photophysical transformations. These findings establish that modulation of pyrene spacing along the peptide backbone and introduction of steric substituents provide a means of tuning spectroscopic responses to pressure. Such scaffold-dependent, multicomponent fluorescent responses under hydrostatic pressurization offer important design principles for the development of ratiometric sensing systems.

6.3 Pressure-Responsive Ratiometric Fluorescence in Pyrene–Polylysine Sensors

Hydrostatic pressure has attracted considerable attention in mechanoscience, and mechanochemistry and mechanobiology have created a strong demand for pressure-responsive chemical sensors capable of detecting minute pressure changes. Against this background, the development of practical ratiometric chemosensors that can be controlled by hydrostatic pressure is strongly motivated.

This section introduces polymer-based chemosensors that emit ratiometric fluorescence signals in response to hydrostatic pressure and remain functional even in buffered solutions relevant to biological applications (Wakako et al., 2023). The design is based on the pyrene fluorophore, which exhibits dual emission originating from both monomer and excimer species, thus being well-suited for ratiometric responses (Fig. 6.7a). α-Poly-L-lysine (α-PLL) and ε-poly-L-lysine (ε-PLL) were employed as polymer backbones. These two scaffolds differ in the number of methylene groups, a variation predicted to influence the polymer dynamics. Furthermore, because the degree of substitution (DS) of pyrene was expected to significantly affect the response characteristics, a series of **α-PLLpyr** and **ε-PLLpyr** with different DS values were synthesized, and a pyrene monomer was used as a reference compound (Fig. 6.7b). These compounds were then compared in methanol, water, and PBS, a buffer solution with ionic strength close to that of biological environments, and their ratiometric behaviors under hydrostatic pressure were investigated. Through these examinations, the influence of solvent environment and DS on the responses was elucidated, with mechanobiological applications being the long-term goal.

Initial characterization at atmospheric pressure revealed that both the α-PLL and ε-PLL systems exhibited enhanced excimer emission with increasing pyrene substitution, reflecting closer spatial packing of chromophores on the polymer backbone. Hydrostatic pressure studies then revealed distinct ratiometric behaviors that depended on both polymer type and substitution level. For example, α-PLL with a high degree of substitution (**α-PLLpyr**$_{\mathbf{high}}$) displayed a decrease in monomer emission intensity accompanied by an increase in excimer emission upon pressurization, yielding a straightforward ratiometric signal (Fig. 6.8a). In contrast, α-PLL with a lower substitution (**α-PLLpyr**$_{\mathbf{mid}}$ and **α-PLLpyr**$_{\mathbf{low}}$) showed simultaneous decreases in both monomer and excimer intensities, although their different pressure dependencies still provided ratiometric information. ε-PLL systems displayed markedly different trends; notably, ε-PLL with medium substitution (**ε-PLLpyr**$_{\mathbf{mid}}$) exhibited the opposite behavior to **α-PLLpyr**$_{\mathbf{high}}$, with monomer emission increasing and

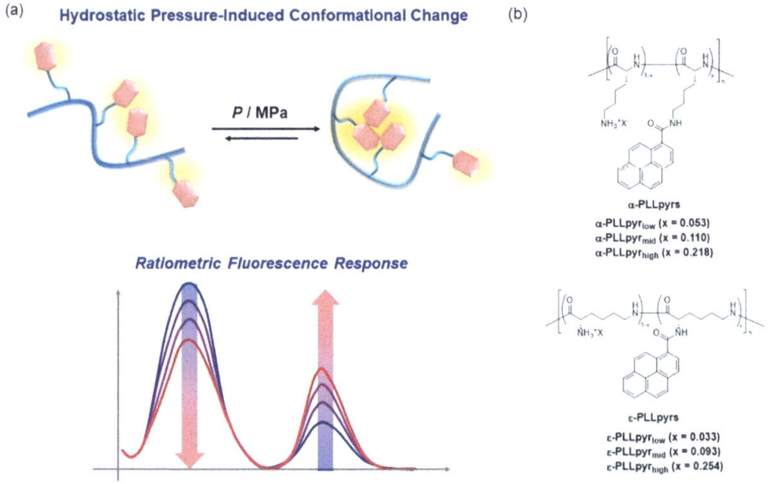

Fig. 6.7 (**a**) Schematic of the ratiometric fluorescence signaling concept driven by hydrostatic pressure-induced conformational changes in polymer chemosensors. (**b**) Chemical structures of the polymer chemosensors (***α***- and ***ε*-PLLpyrs**) and the reference compound **pyr-ref**. (Reproduced with permission. Copyright 2023 American Chemical Society)

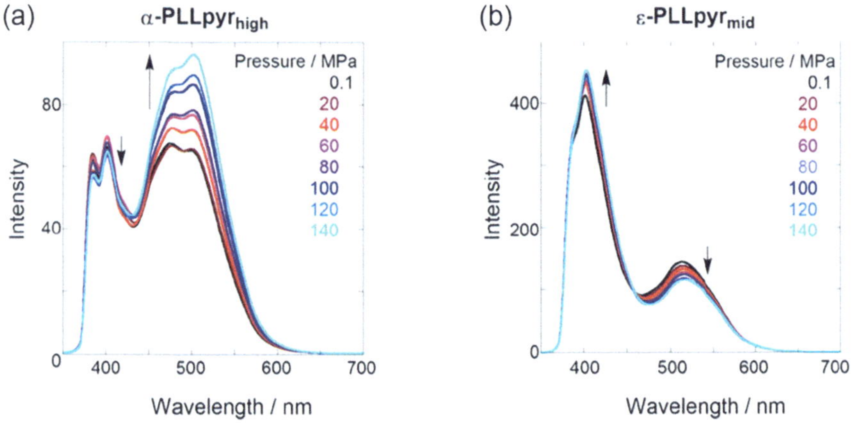

Fig. 6.8 Fluorescence spectra of (**a**) ***α*-PLLpyr$_{high}$** (37 μM) and (**b**) ***ε*-PLLpyr$_{mid}$** (33 μM) in PBS at room temperature, recorded in a high-pressure cell under 0.1–140 MPa (black to sky blue lines). Polymer concentrations are given in chromophore units. (Reproduced with permission. Copyright 2023 American Chemical Society)

excimer emission decreasing under pressure (Fig. 6.8b). These results highlight how subtle differences in polymer backbone flexibility and chromophore density can govern the direction and magnitude of ratiometric pressure responses.

To probe the molecular mechanisms underlying these phenomena, detailed fluorescence lifetime measurements were performed. For **α-PLLpyr$_{high}$**, three distinct emissive species were identified in the monomer region (400 nm), including a short-lived component (τ_1) attributed to proton transfer, a middle species assigned to a ground-state stacked conformer (τ_2), and a long-lived monomeric species (τ_3) (Fig. 6.9). Excimer emission (500 nm) revealed three additional lifetimes, corresponding to different stacking conformations and degrees of overlap between pyrene units. The shortest species (τ_4) was assigned to a ground-state stacked conformation, whereas the two longer-lived species (τ_5 and τ_6) were attributed to excimers formed from alternative stacked conformations. Pressurization reduced both the lifetime and population of the monomer species while enhancing the contribution of sandwich excimers, thereby explaining the observed ratiometric shift. For ε-PLL, similar excited-state species were detected, but pressure reduced excimer-state populations while favoring monomer emission, consistent with the opposite ratiometric trend observed. These findings suggest that the interplay between polymer backbone flexibility, solution viscosity, and ground-state conformer populations dictates the balance between monomer and excimer emission under pressure.

The role of solvent was also examined systematically, because solvent polarity and viscosity significantly influence the conformational dynamics of polymer backbones and the excited-state lifetimes of chromophores. Experiments revealed that ratiometric responses were generally more pronounced in methanol than in water or PBS. This was rationalized by considering both the higher structural order of α-PLL in methanol, as evidenced by CD spectra, and the large pressure dependence of methanol viscosity compared to water. Because viscosity modulates collisional quenching and excimer-formation kinetics, it can strongly influence ratiometric outcomes. In contrast, water and PBS supported more flexible random-coil-like polymer conformations, leading to smaller ratiometric shifts. Overall, these observations demonstrate that the solvent environment is a critical parameter in tuning the magnitude and direction of hydrostatic pressure responses in polymer chemosensors.

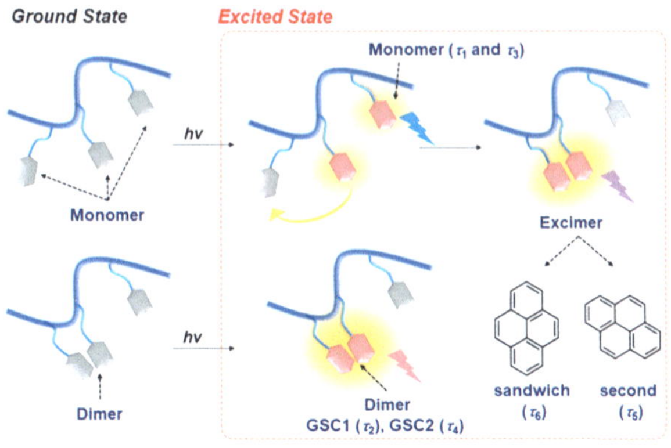

Fig. 6.9 Schematic of the ground-state conformers and excited-state dynamics underlying the hydrostatic pressure-induced ratiometric responses of the chemosensors. (Reproduced with permission. Copyright 2023 American Chemical Society)

To quantify sensor performance, response factors were defined based on the ratio of monomer-to-excimer intensities as a function of applied pressure. These factors varied significantly with polymer type, substitution level, and solvent, ranging from as low as 1.5 to as high as 47×10^{-3} MPa^{-1}. Notably, ε-PLL-based polymers generally exhibited larger response factors than α-PLL-based polymers, highlighting their superior sensitivity as pressure sensors. The ratiometric approach provides clear advantages over conventional on/off fluorescence detection because it minimizes the effects of external fluctuations such as variations in light source intensity or photobleaching, thereby improving the reliability of pressure quantification.

In addition to ratiometric fluorescence, CPL from the pyrene-functionalized polymers was explored, motivated by the chiral nature of the polylysine backbones. Remarkably, **ε-PLLpyr$_{high}$** displayed measurable negative CPL signals in the excimer region, with g_{lum} values of up to 3.2×10^{-3} in water. These values exceed those reported for many chirally-modified pyrene derivatives, and they were attributed to tight chromophore packing on the chiral polymer scaffold. This chiral luminescence introduces an additional ratiometric index, defined as the ratio of left- to right-handed emission intensities, thereby extending the concept of ratiometric sensing into the chiral domain. Such dual functionalities, combining hydrostatic-pressure sensitivity with CPL, represent an innovative step toward multifunctional polymer-based sensors.

In conclusion, the discovery of hydrostatic-pressure-induced ratiometric responses in pyrene-functionalized polylysine polymers represents a significant advance in the development of chemosensors and mechanoresponsive materials. The establishment of reversible pressure detection via robust ratiometric fluorescence provides a powerful platform that is applicable to both fundamental research and practical technologies. By systematically comparing α- and ε-poly-L-lysine scaffolds, varying chromophore substitution levels, and examining solvent environments, key structural and environmental determinants of ratiometric behavior were identified, while fluorescence lifetime analyses revealed the excited-state dynamics governing these responses. Quantitative evaluation of the response factors further confirmed the superior sensitivity of the ε-PLL-based systems. Finally, the observation of CPL activity highlights the feasibility of integrating pressure sensing with chiral optical properties, thereby expanding functional versatility. Collectively, these findings not only deepen the understanding of design principles for mechanoresponsive materials but also open practical avenues for constructing sensitive and selective pressure probes. Such advances hold particular promise for mechanobiological imaging, where precise detection of local pressure variations in cells and tissues is crucial, as well as for advanced materials engineering, where responsive polymers can serve as feedback elements in next-generation smart systems.

6.4 Pressure-Induced Modulation of Supramolecular Chirality in Perylene Bisimide Assemblies

Recent developments in supramolecular chemistry and materials science have shown that functional organic materials can be engineered to undergo structural reorganization when exposed to external stimuli, including temperature, solvent environment, pH, photoexcitation, and hydrostatic pressure. Such responsiveness enables applications in artificial molecular machines, responsive sensors, and switchable systems. Within this broader context, the regulation of chiral phenomena and the control of chiroptical functions have become increasingly important because they provide new approaches to amplify or preserve chirality in dynamic molecular assemblies. Hydrostatic pressure offers a unique route for altering equilibrium structures and kinetic processes. Although earlier investigations focused mainly on thermodynamic equilibria and reaction kinetics, pressure-based approaches are now being revisited in the context of mechanochemistry and mechanobiology, where mechanical forces are recognized as fundamental drivers of structural and functional change. Against this background, clarifying how pressure functions to amplify chirality in dynamic molecular assemblies has become an important objective.

The specific molecular systems chosen for this investigation were perylene bisimide dyes modified with chiral substituents. These dyes are particularly attractive because their chiral supramolecular polymerization properties under ambient conditions have already been well studied, providing a strong baseline for further investigation. Earlier work had also established the chiroptical properties of *N,N′*-bis((*R*)-**4**-phenylethyl)perylene-3,4,9,10-tetracarboxylic bisimide ((*R*,*R*)-**BPP**) and its antipode ((*S*,*S*)-**BPP**), confirming that these molecules form well-defined supramolecular structures in solution and exhibit CD signals that reflect their chiral arrangements (Fig. 6.10) (Taniguchi et al., 2019). Building on this foundation, the study discussed in this section investigated how these materials respond to hydrostatic pressure (0.1–320 MPa) applied in solution (Kinoshita et al., 2021). The aim was to evaluate whether pressure can be used as a controllable stimulus for modulating aggregation behavior and chiral organization, and to evaluate the extent to which pressure-induced changes are detectable by a variety of spectroscopic techniques.

The experimental strategy involved recording UV-vis absorption spectra, fluorescence and excitation spectra, CD spectra, and fluorescence lifetime decay

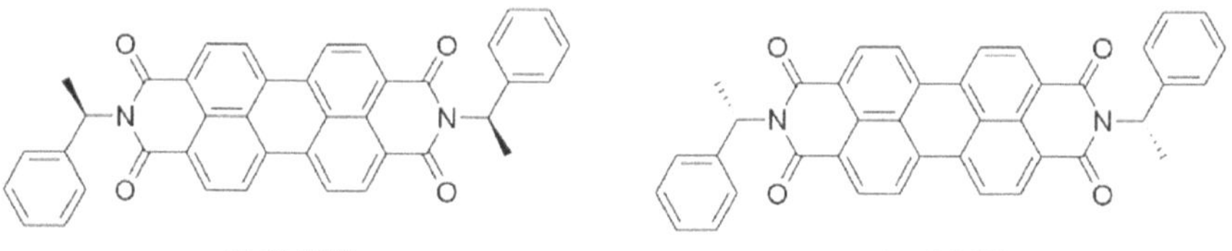

(*R*,*R*)-BPP (*S*,*S*)-BPP

Fig. 6.10 Molecular structures of (*R*,*R*)- and (*S*,*S*)-**BPP**. Reproduced with permission. Copyright 2021 American Chemical Society

profiles of the perylene bisimide solutions under systematically varied pressure. Concentration-dependent UV-vis measurements at ambient pressure were first performed to establish the baseline aggregation behavior (Fig. 6.11a). At higher concentrations, the dye molecules undergo π-stacking interactions that lead to broadened excitation spectra (Fig. 6.11c, d), bathochromic shifts in fluorescence maxima (Fig. 6.11b), and longer fluorescence lifetimes relative to the dilute monomeric form. These features are consistent with *J*-type aggregation, which is typical for perylene bisimide derivatives in halogenated solvents.

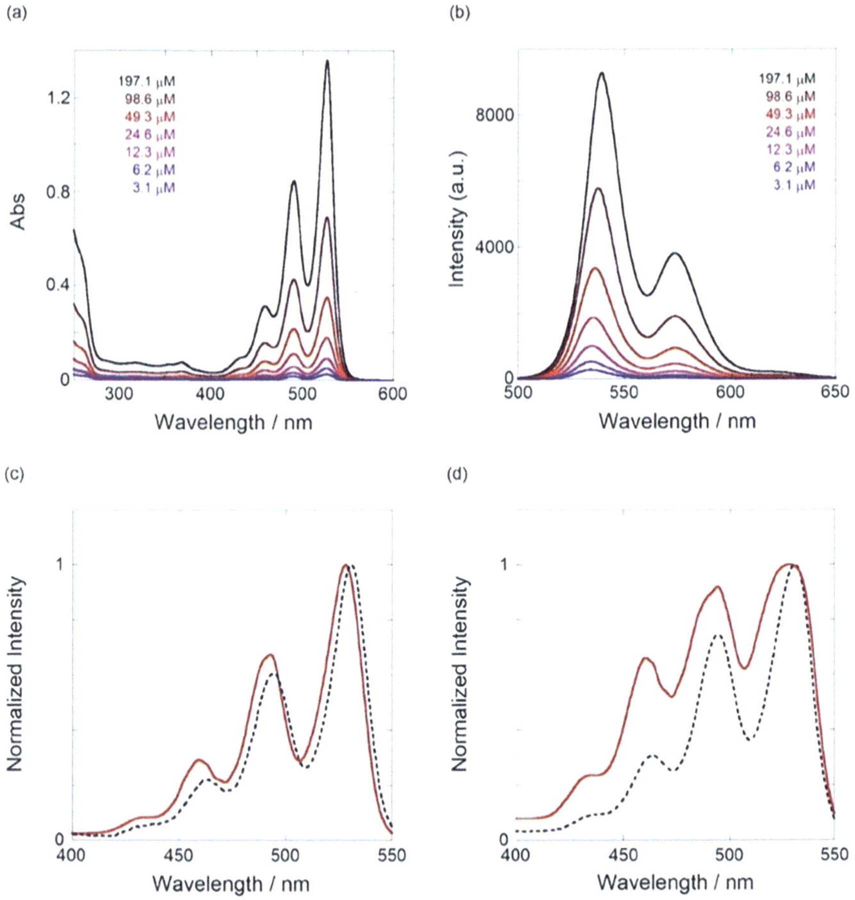

Fig. 6.11 (**a**, **b**) Concentration-dependent (3.1–197.1 μM) (**a**) UV-vis and (**b**) fluorescence (λ_{ex} = 440 nm) spectra of (*S*,*S*)-**BPP** in chloroform at 0.1 MPa and room temperature, recorded using a standard 1 mm cell ((**b**) reflection mode, 55°). (**c**, **d**) Excitation spectra (red lines; (**c**) λ_{em} = 536 nm, 16 μM; (**d**) λ_{em} = 542 nm, 94 μM) and normalized UV-vis spectra (black dotted lines) of (*S*,*S*)-**BPP** at 0.1 MPa in chloroform at room temperature, measured in a high-pressure cell. (Reproduced with permission. Copyright 2021 American Chemical Society)

The spectroscopic properties of the monomeric and aggregated species were evaluated under hydrostatic pressure. When solutions of different concentrations were pressurized up to 320 MPa, both the monomeric and aggregated species exhibited gradual bathochromic shifts in their absorption and fluorescence spectra (Fig. 6.12). Importantly, these shifts were completely reversible upon depressurization, showing that the pressure-induced changes did not damage or permanently alter the molecules but rather modulated their supramolecular organization. Quantitative analysis revealed that the pressure coefficients of the spectral shifts were of similar magnitude for both monomeric and aggregated species, suggesting that the ground- and excited-state conformations are not dramatically altered by pressure at these concentrations. However, the fluorescence intensities decreased stepwise with increasing pressure, indicating that the excited states became increasingly quenched under pressurization. This effect was observed consistently across monomeric and stacked species.

More diagnostic information emerged from the excitation spectra measured under pressure. In monomeric solutions, the excitation spectra retained their sharp features, exhibiting only small pressure-induced shifts (Fig. 6.13a). In concentrated solutions that favored aggregation, the excitation spectra exhibited substantial broadening and more pronounced bathochromic shifts under pressure (Fig. 6.13b). These differences indicate that excitation spectroscopy is a particularly sensitive probe of pressure-dependent changes in aggregate organization, offering a more nuanced picture than conventional UV-vis or steady-state fluorescence measurements. Fluorescence lifetime measurements performed under pressure showed that lifetimes remained consistent with a single exponential decay, suggesting that only one type of excited species was present and that pressure did not induce the formation of new emissive aggregates such as excimers (Fig. 6.14).

The most significant results were obtained from CD measurements under pressure, which allowed the direct monitoring of chiroptical responses in stacked aggregates. Solutions of both enantiomers were prepared at concentrations sufficient to induce chiral aggregation, and their CD spectra were recorded as pressure was increased (Fig. 6.15). In both cases, clear bathochromic shifts were observed, with the extent of the shifts being substantially larger than the corresponding shifts in UV-vis absorption. In fact, the wavenumber shifts in the CD spectra were approximately two to three times greater than those measured in absorption, showing that chiroptical responses are especially sensitive to pressure-induced changes in supramolecular organization. Moreover, the magnitude of the pressure coefficients depended strongly on concentration. For example, at 120 μM the slope of the CD spectral shift was approximately $-1.04\ cm^{-1}\ MPa^{-1}$ at 320 nm (Fig. 6.15a), whereas at 1200 μM, the slope increased dramatically to approximately $-2.13\ cm^{-1}\ MPa^{-1}$ (Fig. 6.15b). This clear distinction provides a valuable metric for using chiroptical signals as indicators of aggregation states in pressure-responsive supramolecular systems.

The enhanced pressure sensitivity of CD likely reflects conformational adjustments around the chiral centers in the perylene bisimide substituents, particularly torsional changes around the C-N bond that propagate through the π-stacked

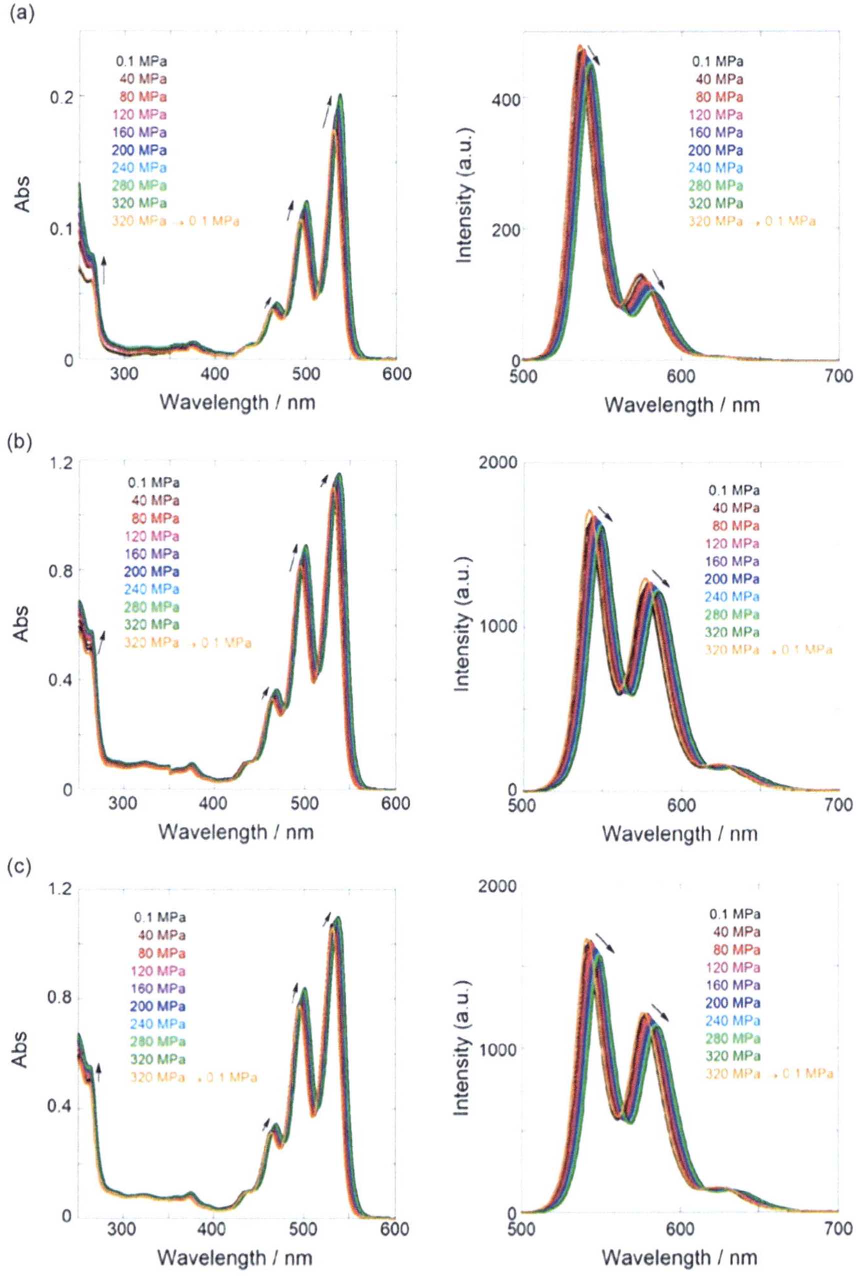

Fig. 6.12 Pressure-dependent UV-vis (left) and fluorescence (right, λ_{ex} = 440 nm) spectra of (**a**) (*S*,*S*)-**BPP** (16 μM), (**b**) (*S*,*S*)-**BPP** (94 μM), and (**c**) (*R*,*R*)-**BPP** (100 μM) in chloroform at room temperature. Spectra were recorded at 0.1, 40, 80, 120, 160, 200, 240, 280, and 320 MPa (from black to green) using a high-pressure cell. The orange trace represents the spectrum at 0.1 MPa after depressurization from 320 MPa. (Reproduced with permission. Copyright 2021 American Chemical Society)

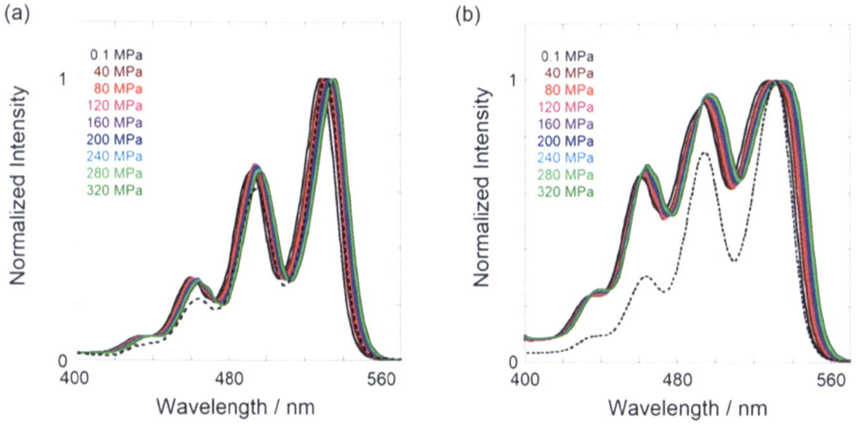

Fig. 6.13 Pressure-dependent excitation spectra (black to green) and normalized UV-vis spectra (black dotted) of (*S*,*S*)-**BPP** ((**a**) λ_{em} = 536 nm, 16 μM; (**b**) λ_{em} = 542 nm, 94 μM) in chloroform at room temperature. Measurements were performed at 0.1, 40, 80, 120, 160, 200, 240, 280, and 320 MPa using a high-pressure cell. (Reproduced with permission. Copyright 2021 American Chemical Society)

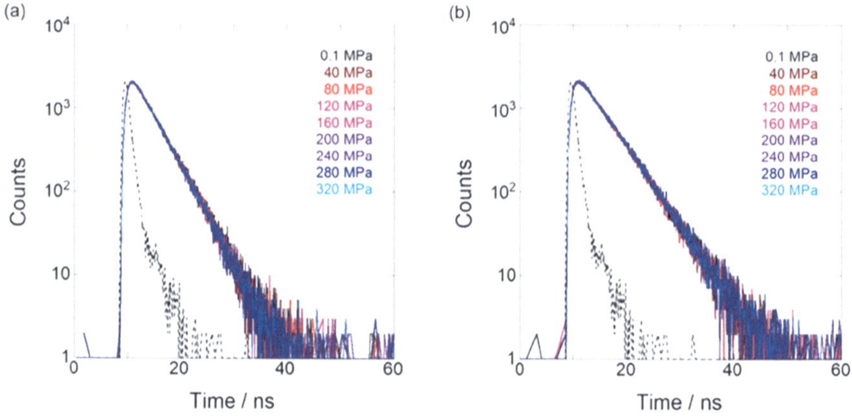

Fig. 6.14 Pressure-dependent time-correlated fluorescence decay profiles of (*S*,*S*)-**BPP** ((**a**) 7 μM; (**b**) 107 μM) in chloroform, monitored at 536 nm under 0.1–320 MPa (0.1, 40, 80, 120, 160, 200, 240, 280, and 320 MPa; from black to sky blue) at room temperature, recorded using a high-pressure cell. The black dotted lines indicate the instrument response function. (Reproduced with permission. Copyright 2021 American Chemical Society)

assembly. As the stacks are compressed, these subtle conformational shifts amplify the overall chiral response, making CD an especially powerful tool for detecting and quantifying pressure-dependent changes in supramolecular chirality. This discovery suggests that hydrostatic pressure can be exploited not only to control chiral aggregation but also to provide analytical insights into the structural dynamics of these assemblies.

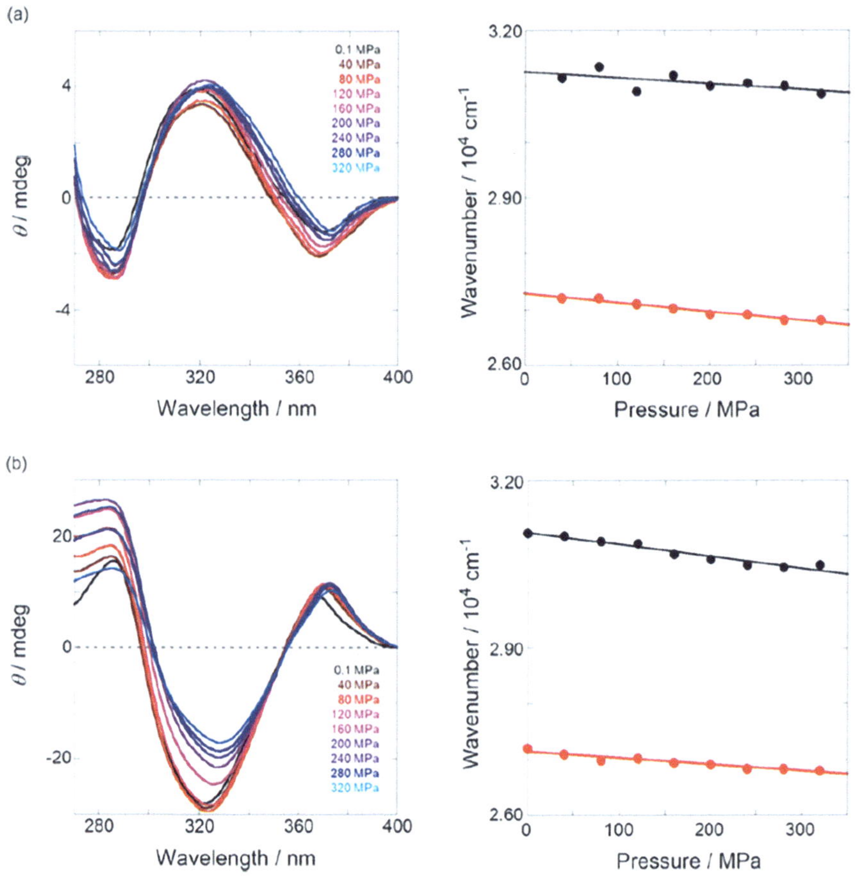

Fig. 6.15 Pressure-dependent CD spectra (left) and corresponding plots of wavenumber shifts at CD maxima (right) for (**a**) (*R*,*R*)-**BPP** (120 μM) (black: $r = 0.639$, slope = -1.04 cm^{-1} MPa^{-1}; red: $r = 0.980$, slope = -1.63 cm^{-1} MPa^{-1}) and (**b**) (*S*,*S*)-**BPP** (1200 μM) (black: $r = 0.971$, slope = -2.13 cm^{-1} MPa^{-1}; red: $r = 0.961$, slope = -1.15 cm^{-1} MPa^{-1}) in chloroform at room temperature. Measurements were performed at 0.1, 40, 80, 120, 160, 200, 240, 280, and 320 MPa using a high-pressure cell. (Reproduced with permission. Copyright 2021 American Chemical Society)

In conclusion, the photophysical properties of chiral perylene bisimide derivatives, both in the monomeric state and in the supramolecular polymeric aggregates, respond to hydrostatic pressure. The UV-vis absorption and fluorescence spectra of both species exhibit comparable pressure-dependent shifts, while excitation spectroscopy more clearly distinguishes the stacked state, as the monomeric species retains sharp peaks that undergo a simple bathochromic shift, whereas the aggregated species shows red shifts accompanied by spectral broadening. Fluorescence lifetime measurements confirm that a single type of emissive species persists under pressurization. Most importantly, CD spectroscopy demonstrates that pressure amplifies chiroptical responses, providing a sensitive and quantitative indicator of supramolecular polymer organization and revealing that subtle conformational rearrangements within the aggregates are reflected in measurable optical signals. These findings establish hydrostatic pressure as a powerful stimulus for both manipulating and analyzing chiral supramolecular polymerizations, offering a versatile approach to study and harness supramolecular chirality. Furthermore, they suggest design strategies for pressure-responsive materials whose chiroptical properties can be tuned in situ, contributing to the development of responsive molecular devices, sensors, and chiral functional systems, and expanding the toolkit of external stimuli for controlling molecular self-assembly and chiroptical function.

References

Cui, Y., Li, B., He, H., Zhou, W., Chen, B., & Qian, G. (2016). Metal–organic frameworks as platforms for functional materials. *Accounts of Chemical Research, 49*, 483–493.

Fukuhara, G., Iida, K., Kawanami, Y., Tanaka, H., Mori, T., & Inoue, Y. (2015). Excited-state dynamics achieved ultimate stereocontrol of photocyclodimerization of anthracenecarboxylates on a glucose scaffold. *Journal of the American Chemical Society, 137*, 15007–15014.

Grabowski, Z. R., Rotkiewicz, K., & Rettig, W. (2003). Structural changes accompanying intramolecular electron transfer: Focus on twisted intramolecular charge-transfer states and structures. *Chemical Reviews, 103*, 3899–4032.

Haidekker, M. A., & Theodorakis, E. A. (2016). Ratiometric mechanosensitive fluorescent dyes: Design and applications. *Journal of Materials Chemistry C, 4*, 2707–2718.

Hawes, C. S., & Gunnlaugsson, T. (2017). Multichannel luminescent lanthanide polymers as ratiometric sensors for D_2O. *Chem, 2*, 463–465.

Karuppannan, S., & Chambron, J.-C. (2011). Supramolecular chemical sensors based on pyrene monomer–excimer dual luminescence. *Chemistry - An Asian Journal, 6*, 964–984.

Kinoshita, T., Imai, Y., & Fukuhara, G. (2021). Hydrostatic pressure-controllable chiroptical properties of chiral perylene bisimide dyes: A chiral aggregation case. *The Journal of Physical Chemistry. B, 125*, 5952–5958.

Lee, M. H., Kim, J. S., & Sessler, J. L. (2015). Small molecule-based ratiometric fluorescence probes for cations, anions, and biomolecules. *Chemical Society Reviews, 44*, 4185–4191.

Luo, J., Xie, Z., Lam, J. W. Y., Cheng, L., Chen, H., Qiu, C., Kwok, H. S., Zhan, X., Liu, Y., Zhu, D., & Tang, B. Z. (2001). Aggregation-induced emission of 1-methyl-1,2,3,4,5-pentaphenylsilole. *Chemical Communications*, 1740–1741.

McLaurin, E. J., Bradshaw, L. R., & Gamelin, D. R. (2013). Dual-emitting nanoscale temperature sensors. *Chemistry of Materials, 25*, 1283–1292.

Mei, J., Leung, N. L. C., Kwok, R. T. K., Lam, J. W. Y., & Tang, B. Z. (2015). Aggregation-induced emission: Together we shine, united we soar! *Chemical Reviews, 115*, 11718–11940.

Mimura, Y., Kitamura, S., Shizuma, M., Kitamatsu, M., & Imai, Y. (2018). Circular dichroism and circularly polarised luminescence of bipyrenyl oligopeptides, with piperidines added in the peptide chains. *Organic & Biomolecular Chemistry, 16*, 8273–8279.

Mizuno, H., Kitamatsu, M., Imai, Y., & Fukuhara, G. (2020). Smart fluorescence materials that are controllable by hydrostatic pressure: Peptide−pyrene conjugates. *ChemPhotoChem, 4*, 502–507.

Nakasha, K., & Fukuhara, G. (2020). Aggregation-induced emission-based polymer materials: Ratiometric fluorescence responses controlled by hydrostatic pressure. *ACS Applied Polymer Materials, 2*, 2303–2310.

Nishikawa, T., Kitamura, S., Kitamatsu, M., Fujiki, M., & Imai, Y. (2016). Peptide magic: Interdistance-sensitive sign inversion of excimer circularly polarized luminescence in Bipyrenyl oligopeptides. *ChemistrySelect, 1*, 831–835.

Padalkar, V. S., & Seki, S. (2016). Excited-state intramolecular proton-transfer (ESIPT)-inspired solid state emitters. *Chemical Society Reviews, 45*, 169–202.

Park, S.-H., Kwon, N., Lee, J.-H., Yoon, J., & Shin, I. (2020). Synthetic ratiometric fluorescent probes for detection of ions. *Chemical Society Reviews, 49*, 143–179.

Sarma, M., & Wong, K.-T. (2018). Exciplex: An intermolecular charge-transfer approach for TADF. *ACS Applied Materials & Interfaces, 10*, 19279–19304.

Sasaki, S., Drummen, G. P. C., & Konishi, G.-I. (2016). Recent advances in twisted intramolecular charge transfer (TICT) fluorescence and related phenomena in materials chemistry. *Journal of Materials Chemistry C, 4*, 2731–2743.

Sedgwick, A. C., Wu, L., Han, H.-H., Bull, S. D., He, X.-P., James, T. D., Sessler, J. L., Tang, B. Z., Tian, H., & Yoon, J. (2018). Excited-state intramolecular proton-transfer (ESIPT) based fluorescence sensors and imaging agents. *Chemical Society Reviews, 47*, 8842–8880.

Taniguchi, A., Kaji, D., Hara, N., Murata, R., Akiyama, S., Harada, T., Sudo, A., Nishikawa, H., & Imai, Y. (2019). Solid-state AIEnh-circularly polarised luminescence of chiral perylene diimide fluorophores. *RSC Advances, 9*, 1976–1981.

Tong, H., Hong, Y., Dong, Y., Häußler, M., Lam, J. W. Y., Li, Z., Guo, Z., Guo, Z., & Tang, B. Z. (2006). Fluorescent "light-up" bioprobes based on tetraphenylethylene derivatives with aggregation-induced emission characteristics. *Chemical Communications*, 3705–3707.

Wakako, S., Hori, Y., Kinoshita, T., Saiki, T., Qi, X., Hasegawa, K., Imai, Y., Mori, T., Nakagawa, K., & Fukuhara, G. (2023). Pressure-responsive polymer chemosensors for hydrostatic-pressure-signal detection: Poly-L-lysine–pyrene conjugates. *ACS Macro Letters, 12*, 1389–1395.

Yanai, N., & Kimizuka, N. (2017). New triplet sensitization routes for photon upconversion: Thermally activated delayed fluorescence molecules, inorganic nanocrystals, and singlet-to-triplet absorption. *Accounts of Chemical Research, 50*, 2487–2495.

Yang, Z., Mao, Z., Xie, Z., Zhang, Y., Liu, S., Zhao, J., Xu, J., Chi, Z., & Aldred, M. P. (2017). Recent advances in organic thermally activated delayed fluorescence materials. *Chemical Society Reviews, 46*, 915–1016.

Yang, D., Han, J., Liu, M., & Duan, P. (2019). Photon upconverted circularly polarized luminescence via triplet–triplet annihilation. *Advanced Materials, 31*, 1805683.

Chapter 7
Conclusion

Abstract The studies presented in this book establish hydrostatic pressure as a quantitative and universally applicable stimulus for probing and controlling molecular systems. By acting isotropically and coupling directly to molecular volume, pressure enables reversible manipulation of ground- and excited-state structures, supramolecular organization, and biological function without altering chemical composition. Across molecular, supramolecular, and biological systems, pressure consistently reveals the intimate interplay among structural flexibility, solvation dynamics, and electronic behavior, allowing free-energy landscapes to be mapped and reshaped through spectroscopic observables. These findings establish pressure not merely as a tool for simulating extreme environments, but as a general actuator capable of translating mechanical constraints into molecular level responses. The demonstrated ability to link volumetric changes with optical signals provides a foundation for pressure-responsive chemosensing, mechanochemistry, and mechanobiology. As high-pressure methodologies and molecular design strategies continue to advance, hydrostatic pressure is poised to play an increasingly central role in the rational control of responsive and soft materials, molecular assemblies, and biological systems.

Keywords Hydrostatic pressure · Soft materials · Mechanochemistry · Mechanobiology

Hydrostatic pressure represents one of the most fundamental thermodynamic parameters governing molecular behavior and chemical reactivity. Acting isotropically in solution and coupling directly to molecular volume, pressure enables perturbation of structure and reorganization energy without altering chemical composition. Through its intrinsic connection to the volume-change term in the Gibbs free energy expression, pressure provides a uniquely quantitative and reversible means of controlling molecular function. These effects allow precise regulation of photochemical reaction processes that are governed by solvation structure and structural relaxation, with the resulting changes readily observable as optical signals.

T. Hamachi, G. Fukuhara, *Hydrostatic Pressure Chemistry*,
https://doi.org/10.1007/978-981-95-7461-2_7

The studies summarized in this book demonstrate that hydrostatic pressure can function not only as an analytical probe of molecular structure and solvation, but also as a universal actuator capable of controlling supramolecular organization, photophysical processes, and biological signaling. Across molecular, supramolecular, polymeric, and biological systems, the application of pressure consistently reveals the intimate coupling among structural flexibility, solvation dynamics, and electronic function. Molecular assemblies exhibiting multiple conformational or electronic states respond to pressurization through redistributions of population that can be quantitatively mapped through spectroscopic observables. These shifts are governed by differences in effective molar volume between states, which integrate intrinsic molecular geometry and solvent reorganization. Therefore, pressure provides a direct experimental handle for exploring and manipulating the free-energy landscape of soft materials. Moreover, chemosensors whose spectral properties respond ratiometrically to pressure offer powerful tools for visualizing subtle pressure variations in materials, living cells, and biological tissues. Consequently, pressure-induced structural control and the accompanying spectral responses offer promising opportunities for advancing mechanochemistry and mechanobiology.

At the molecular level, pressure modifies both ground- and excited-state structures through subtle variations in bond lengths, dihedral angles, solvation shells, and molecular assemblies. Because these effects occur homogeneously throughout a solution, they enable a clear separation of volumetric contributions from other environmental variables such as chemical composition and temperature. In flexible chromophores, such modulation leads to predictable optical responses that reflect the underlying conformational equilibrium. Accordingly, pressure-dependent absorption, emission, and chiroptical signals serve as direct indicators of how molecular geometry and solvent interactions evolve under pressurization. Pressure-responsive chromophores, therefore, provide a mechanistic foundation for designing next-generation pressure-sensitive chemosensors.

In supramolecular systems, pressure offers a powerful means of tuning association-dissociation equilibria. Because molecular recognition involves substantial volumetric changes, modest variations in hydrostatic pressure can dramatically shift equilibrium positions. Pressurization reduces free volume and reorganizes solvation layers, enhancing association when solvent molecules are expelled from cavities or promoting dissociation when internal voids are destabilized. This dual behavior enables reversible and selective control over molecular recognition and self-assembly, thereby establishing a conceptual bridge between mechanochemistry and supramolecular chemistry, demonstrating that macroscopic mechanical stress can be translated into molecular-level reorganization without covalent bond cleavage.

These thermodynamic principles extend naturally to biological and biomimetic systems. Proteins, membranes, and living cells are inherently soft and highly hydrated, rendering them highly sensitive to volumetric perturbations. Pressure alters conformational equilibria, binding affinities, and reaction kinetics, enabling selective interrogation of the forces that stabilize native and nonnative structures. In living cells, moderate hydrostatic pressures modulate intracellular ion fluxes and enzymatic activities in a reversible fashion, revealing that mechanical stress functions as a biochemical signal transducer. These findings underscore that pressure is not only a tool for simulating extreme environments, such as deep-sea conditions, but also a powerful means of elucidating how biological macromolecules and supramolecular assemblies have evolved to balance flexibility and stability under fluctuating mechanical constraints.

In summary, hydrostatic pressure provides a quantitative and universally applicable framework for studying and manipulating molecular systems. By linking volumetric changes to optical responses, this book highlights the universality of pressure as a stimulus that bridges chemistry, materials science, and biology. As high-pressure spectroscopic and microscopic techniques continue to advance, and as the molecular design of functional materials becomes increasingly sophisticated, the integration of pressure control into molecular recognition and reaction coordination is poised to open new directions in responsive materials, molecular machines, and mechanosensing within the realms of mechanochemistry and mechanobiology.

The manufacturer's authorised representative in the EU is Springer
Nature Customer Service Centre GmbH, Europaplatz 3, 69115 Heidelberg,
Germany. If you have any concerns regarding our products, please
contact ProductSafety@springernature.com

Printed and bound by CPI Group (UK) Ltd, Croydon, CR0 4YY
07/10/2026
02229361-0004